RF & Digital Test Equipment You Can Build

Edited by Wayne Green

TAB BOOKS
Blue Ridge Summit, Pa. 17214

FIRST EDITION

FIRST PRINTING—MAY 1974
SECOND PRINTING—NOVEMBER 1976

Printed in the United States
of America

Hardbound Edition: International Standard Book No. 0-8306-4729-5

Paperbound Edition: International Standard Book No. 0-8306-3729-X

Library of Congress Card Number: 74-77022

Preface

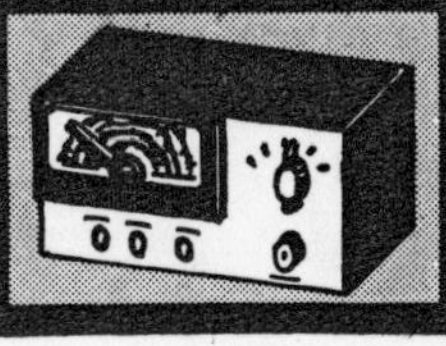

Many of the advanced test instruments used in rf measurements are extremely expensive and complex. It is the purpose of this book to bring the cost of many such instruments within the budget and the theory within the understanding of the typical hobbyist or experimenter.

The projects in this book were originally published in **73 Magazine**, and were selected for this book on the basis of their timeliness and usefulness to the vast majority of electronics readers.

This book is much more than a mere collection of schematics and procedures—though complete schematics and construction procedures for all the instruments covered are given. It represents a rather painless way of acquiring the theory of rf instruments as well as the instruments themselves. We are sure that you will find using the following material profitable, in more ways than one.

The Publishers

Contents

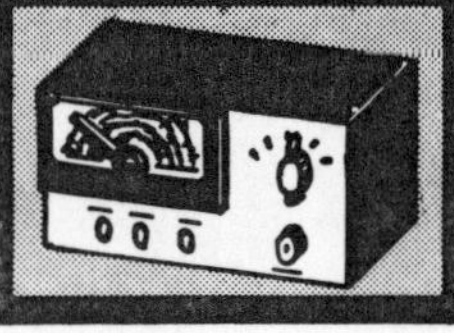

Part I
AF-RF Signal and Waveform Generation

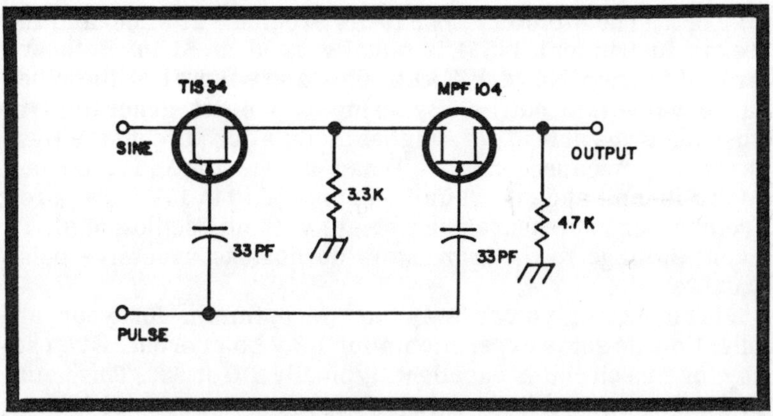

Fig. 6. Final burst generator design.

voltage of the transistor, this circuit may not act as a very good switch anyway—neglecting the biasing. Suppose the sine-wave voltage input is 0.5V and the transistor saturation voltage is 0.2V. We lose a large percentage of our voltage across the transistor.

At any rate, these problems can be completely eliminated by using a junction field-effect transistor (JFET). The JFET measures many megohms between source and drain when cut off, and the **on** resistance (channel resistance) on some of the

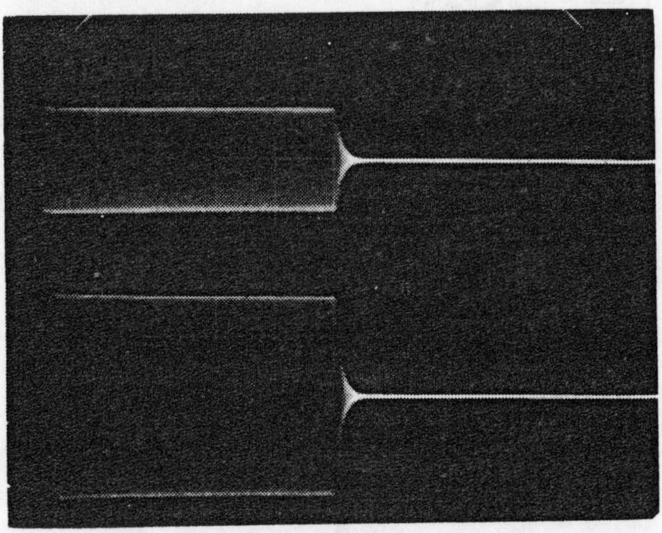

A 1 msec rf burst of 200 kHz, before and after amplification.

good units runs only a few hundred ohms. The FET is used as in Fig. 5. The Motorola MPF104 is excellent at $1.00, and the Texas Instrument TIS34 is equally good at $1.10. Both are available from Newark Electronics and several of the other large wholesale outfits. By using two burst generators in cascade I was able to get a higher modulation percentage (less leakage). Because of what I had at the time, I used one Motorola unit and one TI unit. A pulse of 10 to 15V is required (remember to measure pulse height with an oscilloscope). To avoid damage to the transistors do not use excessive pulse voltage.

The circuit values may not be optimum for your application, so some experimentation may be in order. Even so, R_V for this circuit is excellent, typically 300 or 400. This setup can easily be used from ultrasonic frequencies up to 10 MHz or so.

AF-RF Injector

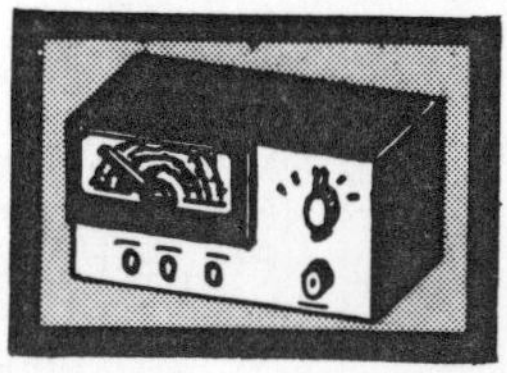

William F. Splichal Jr.

A miniature audio-rf signal generator (Fig. 1) for injecting signals into audio amplifiers or receivers can be constructed with a dozen components and contained in an aluminum cigar tube slightly larger than a pen. The circuit will operate with almost any pair of NPN or PNP transistors, whether silicon or germanium, by merely changing the battery polarity for the type transistors used. Surplus computer circuit boards or junked transistor radios can provide almost all the components necessary to construct this simple generator.

CIRCUIT

The generator circuit, Fig. 2, is basically an astable multivibrator that oscillates at approximately 1000 Hz. The sharp edges of the audio square-wave signal are rich in harmonics up to approximately 200 MHz, and a single 1.5V penlight cell supplies the necessary power. Battery polarity for the various types of small-signal PNP or NPN transistors is as listed in the table.

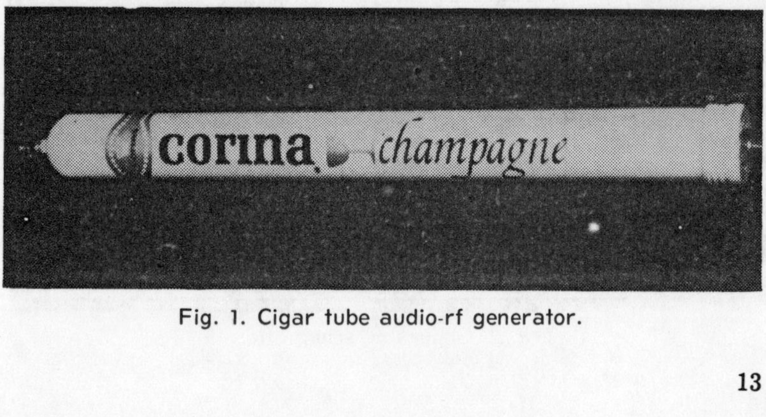

Fig. 1. Cigar tube audio-rf generator.

Transistor Pair	Battery Polarity
2 NPN silicon	As shown on diagram
2 PNP silicon	Reversed
2 NPN germanium	As shown on diagram
2 PNP germanium	Reversed
1 NPN silicon and 1 NPN germanium	As shown on diagram
1 PNP silicon and 1 PNP germanium	Reversed

CONSTRUCTION

An aluminum cigar tube slightly larger in diameter than a penlight cell is used to house the generator. The generator, as shown in Fig. 3, was fabricated on a small piece of Vectorbord and wrapped with electrical insulation tape. The generator may also be constructed on a small printed circuit board or soldered together in a string arrangement and wrapped with insulating tape. A feedthrough insulator with a small pointed tip is attached to one end of the tube to inject the output signal into circuits under test. A small male pin from a scrapped connector could be soldered onto the end of the feedthrough insulator to provide a longer output tip. Layout of the components in the aluminum tube is not critical. A miniature pushbutton switch is located in the other end to apply power to the multivibrator.

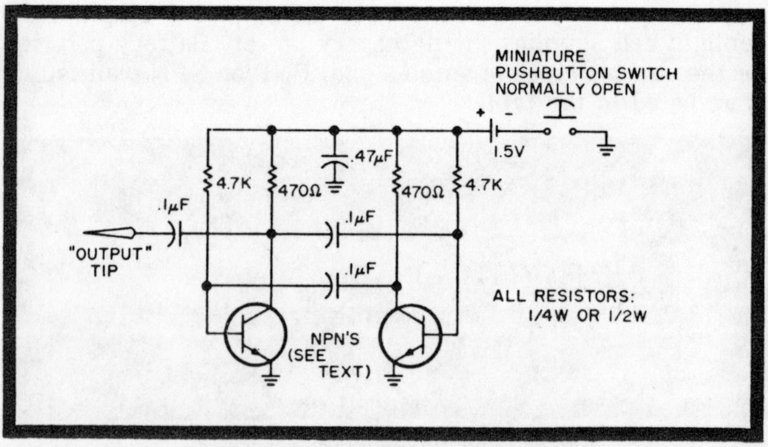

Fig. 2. Generator schematic.

Fig. 3. Component layout.

OPERATION

The signal generator is operated by simply placing the pointed output tip of the generator on the electronic circuit to be tested and depressing the pushbutton switch on the other end. A wire lead may be connected between the case of the aluminum cigar tube and ground of the circuit to be tested. Grounding may also be accomplished by holding the signal generator with one hand and touching ground of the circuit being tested with the other hand. Signal injection into rf circuits may not require the use of a grounding circuit from the generator to the rf circuit under test.

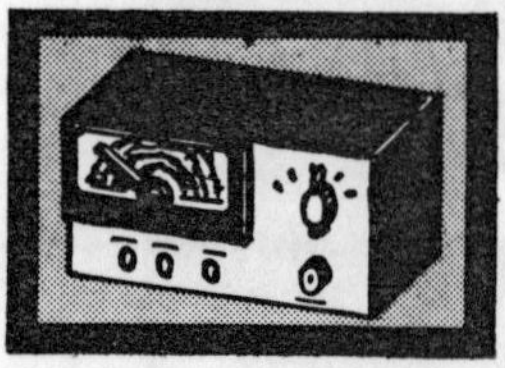

Wien Bridge Oscillator

Parker R. Cope

An audio signal generator is a handy item when you are checking audio gear, and using an af oscillator is a lot more consistent than whistling into the mike. The oscillator described here is the answer. It's inexpensive, simple, and can operate from subaudio to rf.

OPERATION

The oscillator basically is a modified Wien bridge using complementary FET and bipolar transistors. Analysis of the Wien bridge of Fig. 1 shows the bridge is balanced when

$$f = \frac{1}{2\pi\sqrt{R1R2C1C2)}}$$

and

$$C2/C1 = (R3/r) - (R1/R2)$$

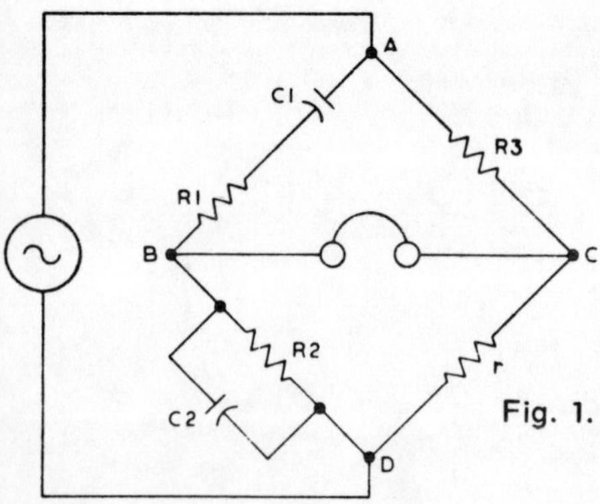

Fig. 1. The basic Wien bridge.

When C1=C2 and R1=R2 and the frequency is such that $R = X_C$, the impedances are

$$Z2 = (R/\sqrt{2})\angle{-45^{\circ}} \quad Z1 = (R\sqrt{2})\angle{-45^{\circ}}$$

Consequently, a third of the voltage applied to the bridge between points A and D appears across Z2. When R3 = 2r, a third of the voltage applied to the bridge appears across r, so there is no difference of potential between points B and C, and the bridge is balanced.

In the actual oscillator shown in Fig. 2, R4 and R5 in parallel make the resistance r (Fig. 1) for ac. The reactance of C3 must be less than 10 ohms at the frequency of oscillation. Resistor R5 is the resistance of a 6W 120V lamp whose positive temperature coefficient regulates the output of the oscillator and insures that signals within the oscillator are not clipped or limited.

A typical 6W 120V lamp has a resistance of 200 ohms when the voltage across it is 0.1V rms, and about 550 ohms when the voltage is 3V rms. A good operating point is about 0.5V, or a resistance of about 300 ohms. The actual operating point is determined by R3 and the particular lamp's characteristics. The larger R3, the higher the lamp operating voltage.

Resistor R3 is chosen to be slightly greater than twice the lamp's resistance when the voltage across the lamp is about 0.5V rms, and a value of 680 is about right. Initially, then, the bridge is unbalanced, and oscillations start when power is

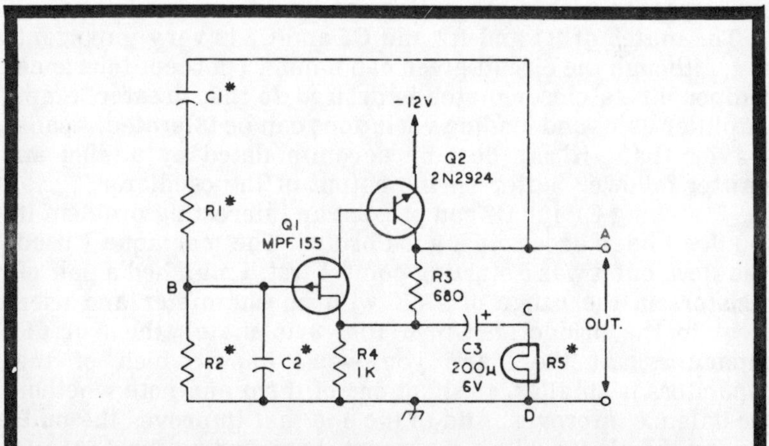

Fig. 2. Fixed frequency oscillator. All resistors 10 percent ¼W.

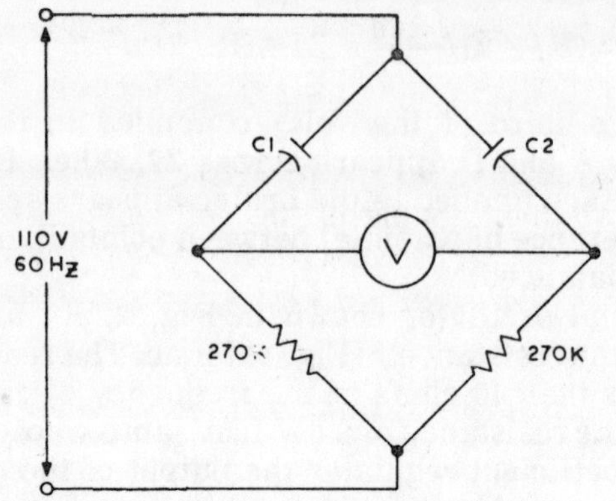

Fig. 3. Capacitor comparison bridge.

applied. The ac voltage across the lamp increases the lamp resistance and brings the bridge toward balance. The bridge is brought to the balance point, which produces a bridge attenuation exactly equal to the amplifier's gain. The attenuation—the ratio of applied voltage (A to D) to output voltage (B to C)—is infinite at true balance, and an infinity-gain amplifier would be required. Therefore, the oscillator must operate with some unbalance.

The match of R1 and R2 and C1 and C2 is very important and, although the circuit given can handle 5 percent tolerance components, a closer match is desired so that greater lamp, amplifier gain, and loading variations can be tolerated. Loads heavier than 1K can best be accommodated by adding an emitter-follower buffer on the output of the oscillator.

Matching C1 and C2 can present an interesting problem if you don't have access to a good bridge. The technique I used was slow, but it was a starting point. First, I matched a pair of resistors in the range of 270K with an ohmmeter and used them in the bridge shown in Fig. 3 to match the 0.01 uF capacitors. At the start, you won't know which of the capacitors is smaller, so shunt one of them and note whether the balance improves. Add to the one that improves the null, and try to balance within 2 percent. In matching the 0.001 uF one, the bridge resistors should be on the order of 2.7M, and a VTVM should be used for detecting the null.

Values of R1 and R2 between 2K and 1M and values of C1 and C2 above 470 pF are convenient. Within these bounds, the combinations in Fig. 4 are available.

The same general circuit approach can be used to make a wide-range variable frequency oscillator. In the variable oscillator, fixed resistors R1 and R2 are replaced with sections of a dual pot. If the tracking of the pot sections is within 5 percent, the circuit can be used directly, and a 100:1 tuning range can be achieved, but it isn't very likely that you will find such close tracking.

There are two possible solutions to the difficulty. One reduces the tuning range by adding fixed resistance in series with the variable resistances to reduce the percentage tracking error, and the other increases the gain of the amplifier so the unavoidable unbalance can be tolerated. In the oscillator shown in Fig. 5, both solutions are used. Rab adds about 2.5K in series with each of the 100K pot sections and reduces the tuning range, and Q3 makes a higher voltage gain possible from Q2. The oscillator tunes from about 150 Hz to 6 kHz for the values given.

The pot R_{ab} is adjusted to minimize the percentage difference between R1 + R_a and R2 + R_b as the dual pot is rotated fully clockwise. The power supply for the oscillator is not critical, and any voltage from 12V to 24V will do for the fixed-frequency oscillator should have a supply above 18V. The current required for the fixed oscillator is about 5 mA without a buffer, and the variable oscillator requires about 18 mA.

Freq. (kHz)	C1, C2 (uF)	R1, R2, (k)
0.250	0.01	63.4
0.400	0.01	39.2
1.0	0.01	15.8
1.8	0.001	88.7
2.5	0.001	63.4
10	0.001	15.8

Fig. 4. Component combinations.

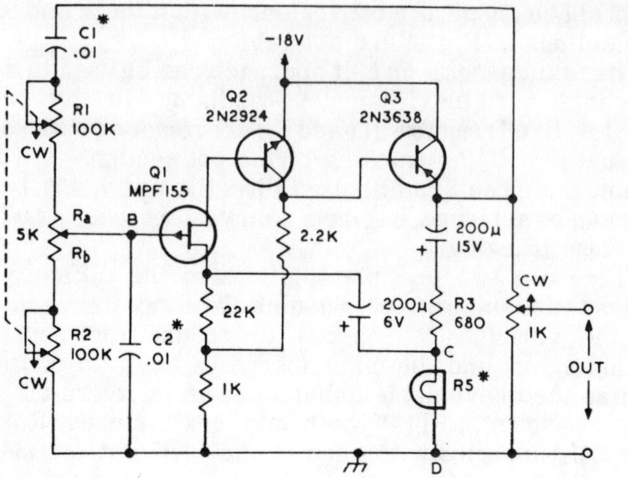

Fixed Osc.	Emitter (Source)	Base (Gate)	Collector (Drain)
Q1	—4.5	0	—11.4
Q2	—12V	11.4	—7.5
Var. Osc			
Q1	—4.5	0	—17.4
Q2	—18.	—17.4	—14.4
Q3	—18.8	—14.4	—18

Fig. 5. Variable frequency oscillator. All resistors are 10 percent ¼ W.

CONSTRUCTION

The layout of the oscillator is not particularly critical for audio frequencies, but care should be taken to shield the high-impedance sections from power-line pickup. If an external power supply or batteries are used, the normal enclosure will be sufficient. The variable oscillator of Fig. 4 will fit in a minibox if an external power supply is used. The heft will be improved if you bolt a chunk of scrap steel into the bottom of the box.

If you prefer, you can change the polarity of the unit by making these direct substitutions. Change MPF 155 to MPF 104, 2N2925 to MPF 6518, and 2N3638 to 2N2923. Don't forget to change the polarity of the electrolytics. The heavy negative feedback makes the circuit very tolerant of component variations, but the table of dc voltages may be comforting when you turn the oscillator on for the first time.

An evening or two and a few bucks for parts are all that are needed to build the oscillator. If your junkbox is like mine, full of everything except what you need, it will cost you under $6 to buy all new parts for the fixed oscillator.

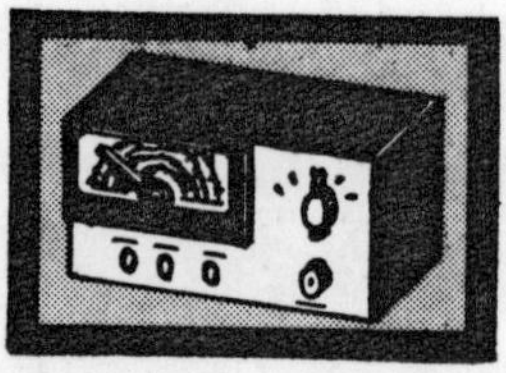

Waveform Generator

Richard Factor

Most experimenters have audio frequency oscillators, yet their use is severely limited by the lower frequency limit of about 20 Hz in most low-cost units. This limitation exists because of the extremely high impedance necessary to avoid loading the frequency-determining network. Until the advent of the field-effect transistor (FET) it was almost impossible to build a low-distortion, low-frequency transistor signal generator. It is still difficult to lower the output frequency significantly without introducing severe distortion.

The unit described in this article has none of the low-frequency limitations of bridge-type oscillators. It generates signals in the range of 4 kHz down to about 10 mHz (that's 10 **millihertz**, or 0.01 cycle per second!). Such frequencies have many uses in generating slow-sweep displays, checking the response of low-frequency filters, voltage-control applications in electronic music generation, instrumentation, etc. The circuitry is similar to that of a high-priced laboratory instrument known as a **function generator**, but by taking advantage of low-cost ICs and nonprecision resistors, it can be built for substantially less than most kit-type audio generators.

CIRCUITRY

Most audio oscillators directly generate a sine wave. Sine-wave oscillators usually are much more critical than other types. The oscillator of Fig. 1 generates a triangular wave, which is then shaped into a sine wave using a FET instead of a reactive filter. This renders it insensitive to frequency, enabling the distortion to be constant over the entire range. Due to difficulties in obtaining precise symmetry in FET characteristics, plus nonideal characteristics of the transfer function itself, it is difficult to get the distortion lower than about 1.2 percent. This is adequate for almost all applications.

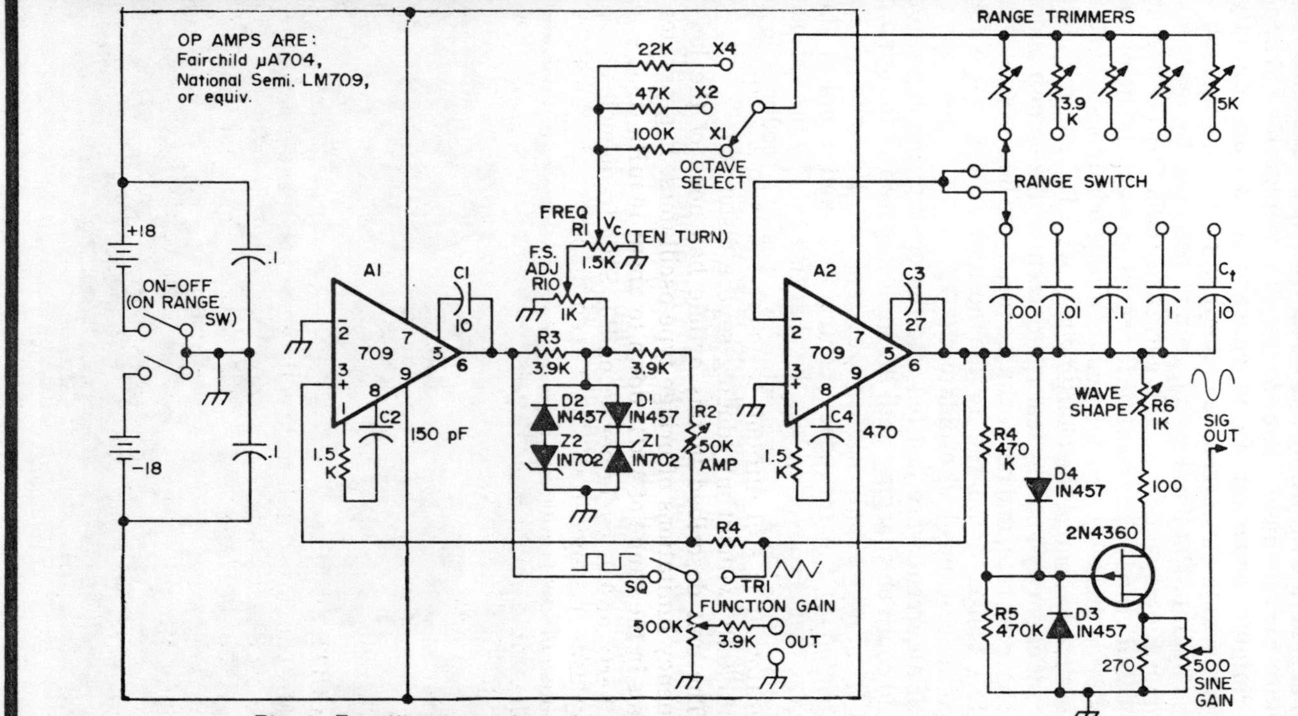

Fig. 1. Function generator schematic diagram.

Integrated-circuit operational amplifiers (opamps) are used to generate the triangular wave. Unit A2 is connected in a circuit arrangement known as an integrator. Time-determining capacitor C_t is charged at a rate directly proportional to the current fed into pin 2 (the inverting input) of the opamp. Pin 2 is kept at ground potential by feedback through the capacitor, so the current equals the voltage at point V_c divided by the total series resistance consisting of the resistor on the octave switch and the range trimmers. Thus, it is possible to control the charging rate with R1, and the high current capacity and low leakage current of the opamp allow the charging rate, and hence the frequency, to be varied over a 500-to-1 range, compared to the normal 10-to-1 range for capacitance-controlled oscillators.

If a current were just fed into the opamp, the capacitor would continue charging until the limit imposed by the opamp and the power supply voltage was reached. However, A1 is connected to the frequency control pot, and supplies the voltage that is converted into the charging rate by the various resistors. Opamp A1 is connected in a positive feedback loop with high hysteresis. The opamp is sensitive to the voltage between the two input terminals, and if the difference is only a millivolt or so, the amplifier goes to either its fully positive or fully negative limit, depending on the input polarity.

Feedback to the input is through R2 and R4. Assume that the output of A1 is fully negative. This voltage is clamped to

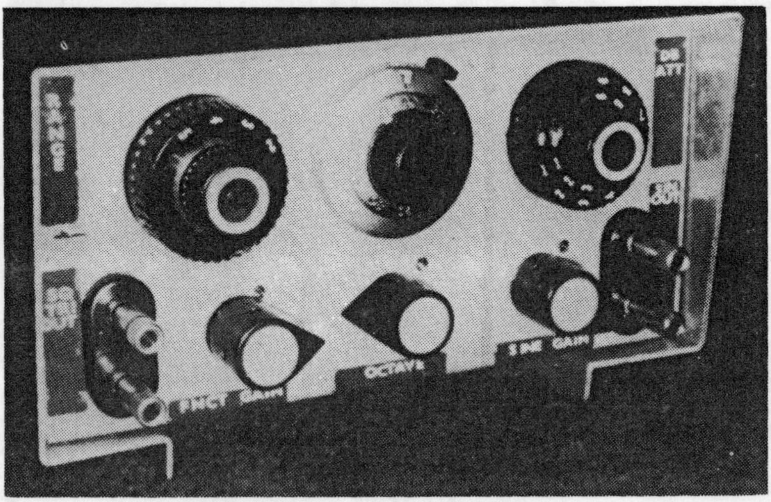

Front panel of function generator.

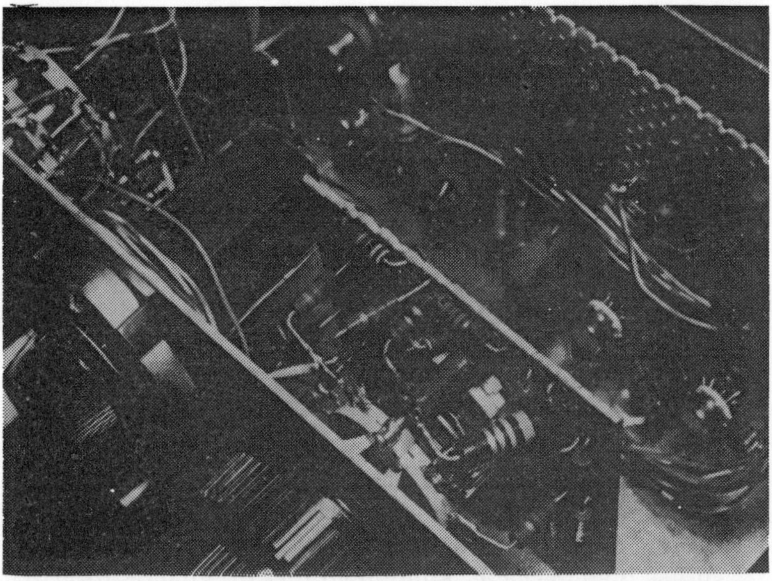
Interior view of unit with step attenuator resistors in front.

about 4V by D2 and Z2 (D1 and Z1, when positive). This 4V is connected back to the input through R2 and the 3.9K protective resistor. As the voltage at the output of A2 gets increasingly positive, a point is reached where the current through R4 becomes opposite to and slightly greater than that through R2. When this happens, the amplifier almost instantly (in a few microseconds) changes state, aided by the positive feedback through R2. Of course, when the opamp changes state, the integrator starts to charge in the opposite direction, thus creating the triangular wave. The zener diodes insure that the wave is highly symmetrical. If your zeners are out of tolerance, or distortion is critical, connecting a small resistor in one leg or the other might help a little.

The sine wave, square wave, and triangle wave are all synchronous and available simultaneously. To conserve panel space on my unit, I used a switch to connect either the square or the triangle to the binding post, though this is not necessary. Whatever arrangement you use, avoid loading the output on the opamp by more than about 2K. It won't harm anything, but the output frequency may change slightly when connecting and disconnecting loads if this precaution isn't observed. It is normally very difficult to measure the frequency of VLF sine waves. Connecting the square-wave output to a counter while using the sine output is a great convenience.

The sine-wave shaper takes advantage of the symmetry of the source and drain with respect to the gate of the 2N4360 JFET. The 2N4360 was chosen primarily for its very low cost, as well as its relatively high pinchoff voltage, which gives a greater output voltage. The output of A2 varies symmetrically about ground, and as the absolute value of the voltage increases, the FET resistance also increases due to the gate bias. Incidentally, note that the power to drive the FET circuit is supplied directly by the opamp output. Proper bias polarity is insured by D3 and D4.

Since the output is symmetrical about ground, no coupling capacitors are necessary. It would be hopelessly unwieldy to use one for 0.01 Hz anyway. The output amplifier is designed for unity gain with no voltage offset. Simplicity is achieved through the use of complementary transistors. The output impedance is under 100 ohms. Resistors R6, R7, R8, R9, and diodes D5 and D6 are the bias network, which eliminates crossover distortion in the amplifier.

If you plan to use the equipment for checking professional audio equipment, you might desire to add the attenuator shown in Fig. 2. It provides 89 dB of attenuation into a 600-ohm load. Two concentric single-pole, 10-position switches break the 89 dB into 1 dB steps.

To simplify the attenuator design and allow the use of standard resistors with 5 percent tolerance, an unusual circuit was used. The 10 dB steps are provided by a constant impedance (200-ohm) T-attenuator, and the single dB steps are provided by switching in series resistors. Thus, the attenuator is not of a constant impedance design, although the voltage into a 600-ohm load conforms to the proper setting.

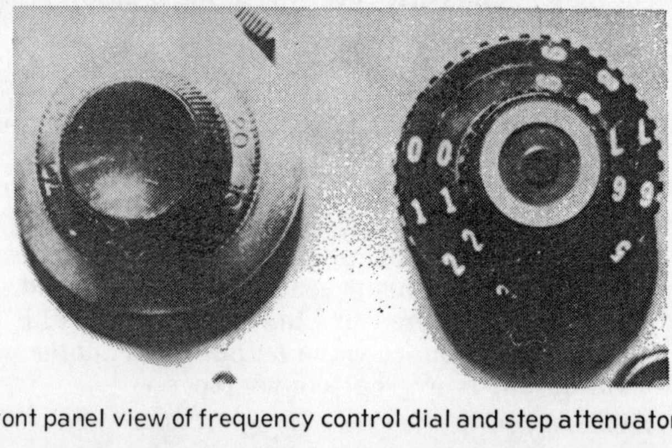

Front panel view of frequency control dial and step attenuator dial.

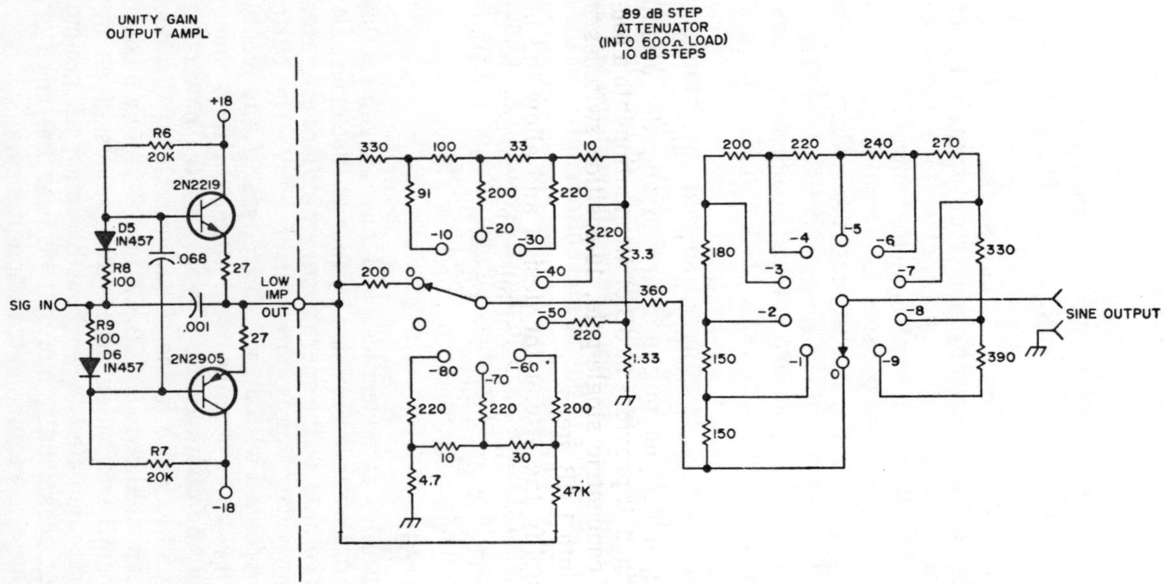

Fig. 2. Output amplifier and attenuator schematic.

If you are using a high-impedance load, the output should be bridged with a 600-ohm resistor. A low-impedance load will exaggerate the effect of the 1 dB step switch. I emphasize that this whole assembly, while desirable, is purely optional.

ADJUSTMENT

As you can see from the photograph, the frequency control is a 100-turn precision potentiometer. These are readily available for a couple of dollars in the surplus market. Using one, it is possible to calibrate the unit so that, for the top 900 divisions, the output frequency is correct to within 1 percent. As the **octave-select** switch is switched to a lower series resistance, the linearity of the **frequency** control is decreased. The **full-scale adjustment** pot sets the output frequency to 100 Hz on the range with the 0.01 uF capacitor when the **octave** switch is in the **x1** position (100K series resistance). Then, the various **range-adjust** pots are set until each range has the appropriate full-scale value: 1 kHz, 100 Hz, 10 Hz, 1 Hz, and 0.1 Hz.

Potentiometer R2 is adjusted so that the triangle amplitude is proper for conversion to a sine. This control interacts with the **full-scale adjustment** pot, so they both may have to be readjusted several times. **Waveshape** pot R6 is adjusted for symmetry in the top and bottom halves of the sine wave. A scope or harmonic distortion meter is advisable for the above two adjustments, although a reasonable null can be achieved by a musically trained ear, especially if there is an undistorted signal available for comparison.

CONSTRUCTION

The circuitry is noncritical and any convenient layout may be used. Power is provided by four 9V batteries. Do not exceed this voltage, as the operational amplifiers are rated for a maximum power supply potential of ±18V. The batteries should be bypassed with 0.1 uF capacitors. If any trouble is encountered with the opamps oscillating at high frequency, it will probably be eliminated by bypassing the power supply pins to ground as close to the IC as possible. If this doesn't work, add capacitance to C1, C2, C3, and C4, a few picofarads at a time until oscillation stops. The values given in the diagram are calculated to work with almost all production opamps, but occasionally one will be near the edge of the specifications and require extra capacitance.

If you don't need precise frequency calibration or attenuation, substantial cost can be saved by eliminating these

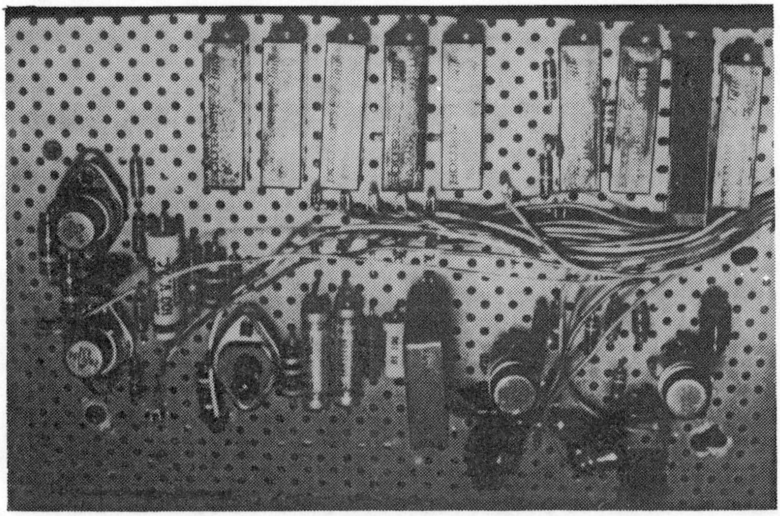

Closeup view of circuit board.

features. The range trimmers can also be eliminated (just replace with a short circuit). The **octave** switch can be replaced with a fixed resistor down to 15K or so.

The output attenuator can be replaced with an ordinary pot. Substitute the 500-ohm pot and 470-ohm opamp resistor before the amplifier with a 240-ohm resistor; connect the opamp input directly to the FET. The pot must be connected to the amplifier output to attenuate the amplifier noise as well as the signal.

Use any convenient pot for the frequency control, and calibrate the dial yourself.

Following these procedures can reduce the cost to under $10 and yet provide you with an oscillator that, in the lower frequency ranges, is better than anything selling for under the price of a laboratory function generator.

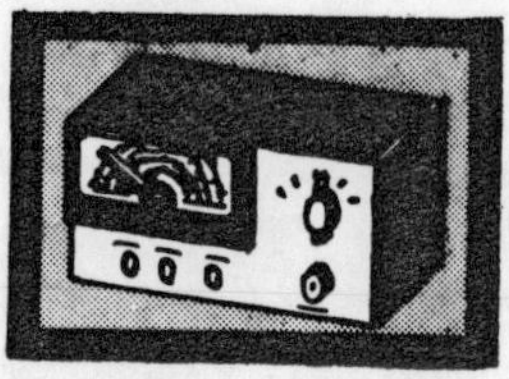

IC Square-Wave Generator

Charles Jimenez

This project was designed primarily for those who wish to acquaint themselves with, and gain experience using, integrated circuits. The square-wave generator described is a rather easy construction project. This is not to say, however, that it sacrifices performance for the sake of simplicity. In fact, several shortcomings of usual square-wave generator circuits have been overcome in this design.

The integrated-circuit square-wave generator.

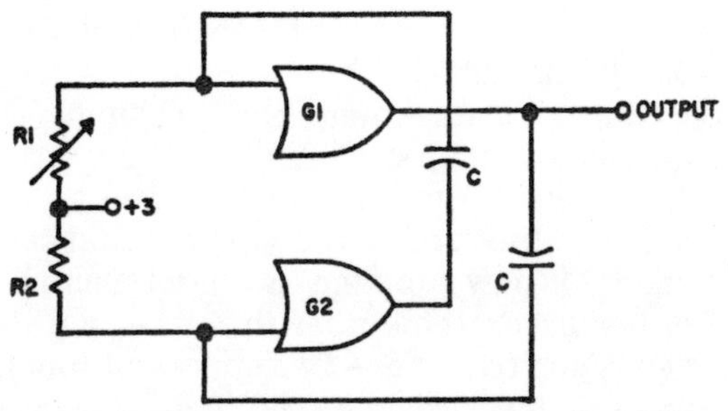

Fig. 1. Simple astable multivibrator circuits.

CIRCUIT DETAILS

The circuit that generates the basic square waveform is shown in Fig. 1. This is called an astable multivibrator. Gates G1 and G2 are from a Fairchild uL 914 integrated circuit. It's a dual two-input gate and should be familiar to many readers.

Each gate is cross-coupled to the other through a resistor-capacitor network, which determines the operating frequency. Different capacitors are switched in for changing frequency bands. There are five bands: band A, 10 Hz to 150 Hz; band B, 100 Hz to 1.5 kHz; band C, 1 kHz to 15 kHz; band D, 10 kHz to 150 kHz; and band E, 70 kHz to 1 MHz. In order to vary the frequency within these bands, normally you have to vary both R1 and R2 simultaneously. But by varying only R1, we can obtain the same bandspread and save the cost of a ganged pot. Unfortunately this will destroy the symmetry, or squareness, of the output waveform. This can be remedied and, as you will see later, the remedy brings along a couple of extra advantages of its own.

The simple astable multivibrator of Fig. 1 would work nicely if it were not for one big shortcoming. It may cease oscillating when switching frequency ranges or it may fail to start up when turned on. This happens when both gates saturate at the same time. In normal operation, G1 and G2 conduct on alternate cycles; that is, when G1 conducts, G2 is cut off. This process is insured by the capacitors, which drive the gates by charging and discharging alternately.

However, suppose now that you are changing bands. As the arm of the bandswitch moves from one capacitor to the next there will be a time interval when there is no capacitor in the circuit at all. Both gates will now see a positive voltage at their inputs through G1 and R2 and will conduct heavily. The multivibrator will now be locked and cannot be started up again unless you first turn off the power. Obviously, it would be very frustrating to have to turn off the power whenever you wanted to change bands.

The seemingly insurmountable problem was easily overcome by using a couple of diodes. Figure 2 shows the circuit, known as a self-starting circuit. The two diodes are connected to each output and to the junction marked "X". The +3V for R1 and R2 is now supplied through D1 and D2 from the collector of either gate. Remember that when a gate is cut off the collector goes positive and +3V appears at junction X. The

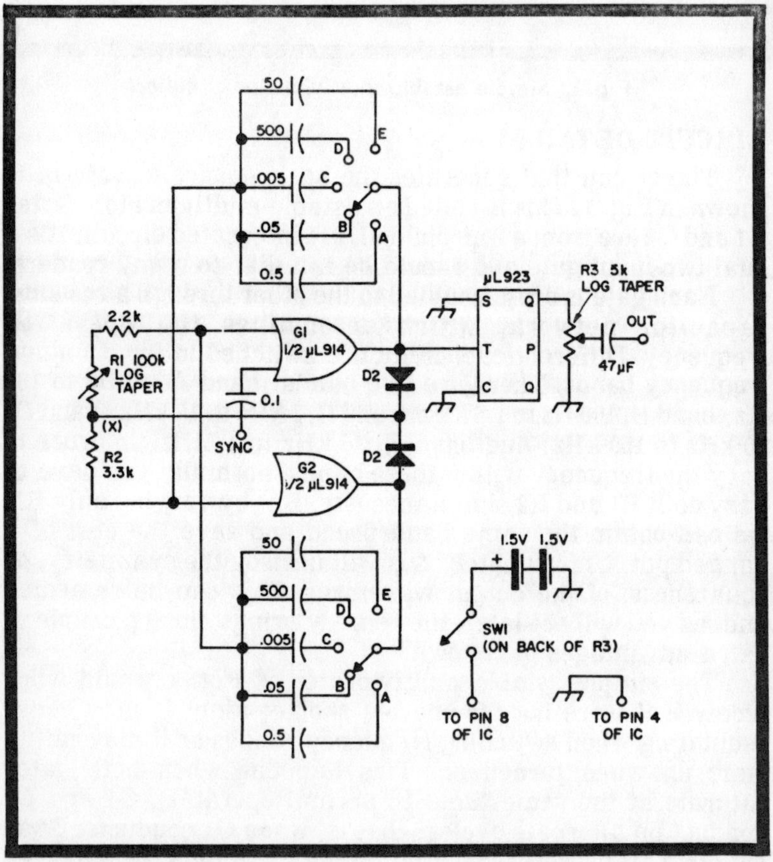

Fig. 2. Complete square-wave generator. Bandswitching capacitors are 10 percent, or better, tolerance. Resistors are ¼W.

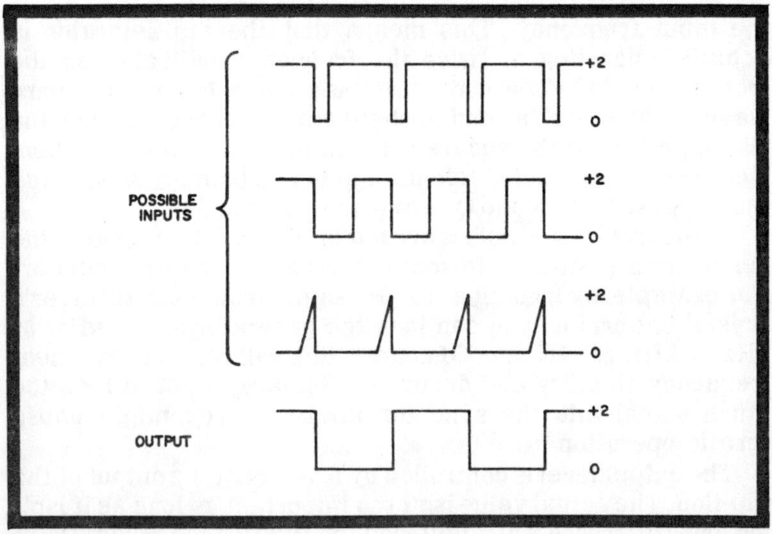

Fig. 3. Operation of the uL 923 flip-flop. Note that regardless of the shape of the input waveform, the output is always a perfectly symmetrical square wave.

circuit will operate properly as long as at least one gate is cut off. Now if both gates should happen to saturate at the same time when switching capacitors, the positive voltage at X will disappear, tending to cut off the gates immediately. In other words, the diodes, which form the OR gate, will not allow the multivibrator to lock in a saturated condition. Proper operation will begin when the next capacitor is switched in. We now have a reliable astable multivibrator circuit, which produces rectangular waves.

As stated earlier, the method used for varying the frequency destroys the output wave's symmetry. When R1 is varied, the output may change from a square wave to a rectangular wave or pulse, for instance. Of course, this change in waveshape has no effect upon the frequency as it is varied. In order to correct the waveshape, the output of the multivibrator is fed into a Fairchild uL 923 J-K flip-flop.

The action of this flip-flop is shown in Fig. 3. The uL 923 output changes only when the input signal goes negative. Notice that the output is always a perfectly symmetrical square wave, regardless of the shape of the input waveform. The input can be spikes, pulses, rectangular waves, or any other waveform that has a fast negative-going portion. It can also be seen from the diagram that the output frequency is half the input frequency. This means that the multivibrator is actually operating at twice the frequency indicated on the front-panel dial. The generator puts out a beautiful square wave to 1 MHz and beyond. A slight amount of overshoot on the rising portion of the square wave is normal at high frequencies. The S (set) and C (clear) inputs are both grounded, and the P (preset) input and $\overline{Q}$ output are disregarded.

Synchronization pulses are fed into gate G1 for locking the generator's frequency to some external source or oscillator. For example, by feeding a 100 kHz signal from your receiver's crystal calibrator, you can lock the generator at 100 kHz, 50 kHz, 33 kHz, 25 kHz, etc. Of course, this will result in excellent frequency stability and accuracy. Be careful not to feed too much signal into the sync terminals, as you might cause erratic operation.

The output level is controlled by a 5K pot at the output of the flip-flop. The actual value isn't too important as long as it isn't too low. Otherwise you might load the flip-flop too much. Don't go below 1K. The use of a log-taper pot will permit adjustment down into the millivolt region for low-level audio work. The output voltage is about 2V into a high-impedance load.

The supply voltage for the unit is taken from two 1.5V D-cells in series. Current drain is less than 40 mA. Remember that pin 8 of both ICs is connected to the +3V; and pin 4 of both is connected to ground, or minus. A colored line or flat portion on the edge of the IC body identified pin 8. Diodes D1 and D2 can be almost any signal diode. Parts values should be followed rather closely to insure adequate band coverage.

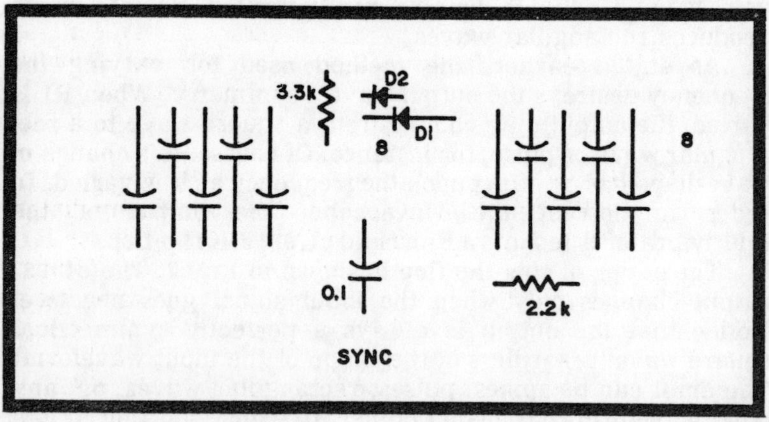

Fig. 4. Full sized layout of PC board. This is a bottom view, with components mounted on top.

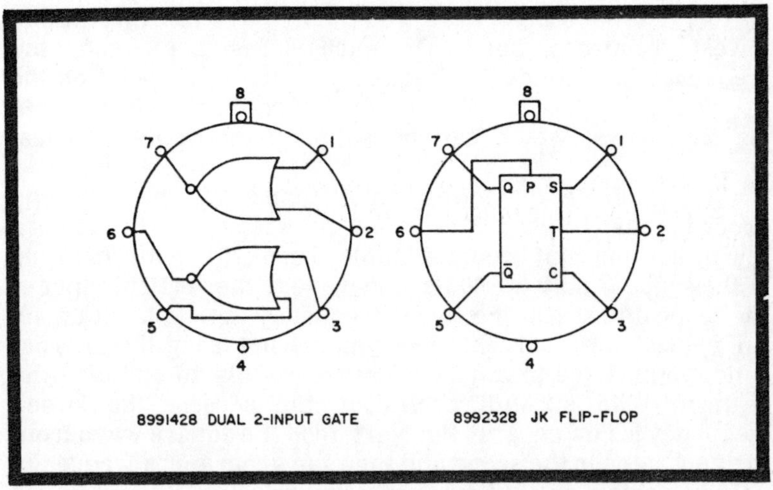

Fig. 5. Basic diagram for the ICs.

CONSTRUCTION

The printed-circuit layout (Fig. 4) is given for those who want to make their own PC boards. You can get an idea of the front-panel arrangement from the photo. Actually, there is nothing critical about layout or construction, so you can arrange things inside to your liking. I used a 4 x 5 x 6 in. minibox for my unit, which is just right, if you use a Millen 10039 vernier dial as I did. This is a compact unit, and using a larger dial will mean using a larger cabinet. The Millen dial is rather expensive and maybe you'll want to use one of the imports and make your own scale. Since I'm on the subject of cost, I might as well say that the whole project will come to about $20 with all new parts, including the Millen dial. With an imported dial, you can probably knock $5 off that figure.

CALIBRATION

You might have noticed by now that the scale on my dial is not linear. This is because I used a linear taper pot for R1, since it was available. I'm not particular about such things, but if you prefer a more linear scale, I would suggest trying a log or semilog taper pot. Keep in mind the fact that most vernier dials turn only 180 degrees as opposed to the normal 270-degree turn of a pot. You might have to adjust the position of the pot in the dial to insure proper bandspread.

Calibration can be achieved only by the use of a scope or frequency counter. If you don't own one, maybe you can gain access to one for about 15 min or so. By using 10 percent or better, tolerance capacitors, you'll only have to calibrate the lowest frequency band. On each succeeding band, the frequency is 10 times the frequency at the same point on the previous band. The simplest method of calibration is with a 60 Hz sine wave, which can be supplied internally on most scopes.

To calibrate the lowest band (band A), the following procedure can be used. First, allow the scope to warm up for a few minutes until it becomes stable. Turn off the internal sync of the scope. Apply the 60 Hz sine wave to the vertical input of the scope and adjust the sweep frequency until you obtain six full cycles on the screen. Since you are not using the internal sync, you'll have to adjust it very carefully to stabilize the pattern. With six full cycles on the screen, the sweep frequency is now set at 10 Hz. Next, feed the square wave from the generator to the scope and tune the generator's frequency until you obtain one full cycle of a square wave. Be careful not to move the sweep frequency of the scope. The square-wave generator is now set at 10 Hz and can be marked on the dial. Tune the generator again until two full cycles are visible on the screen. The scale can now be marked at 20 Hz. This process can be continued on up to 100 Hz. Afterwards, go back and repeat it to make sure you have the proper calibration. Once you have made certain that there are no errors, you can mark the rest of the bands as outlined previously. On band E, you can listen to the signal on a broadcast receiver to see if it checks out. The bands on my unit did not come out exactly in multiples of ten because I used 20-percent tolerance

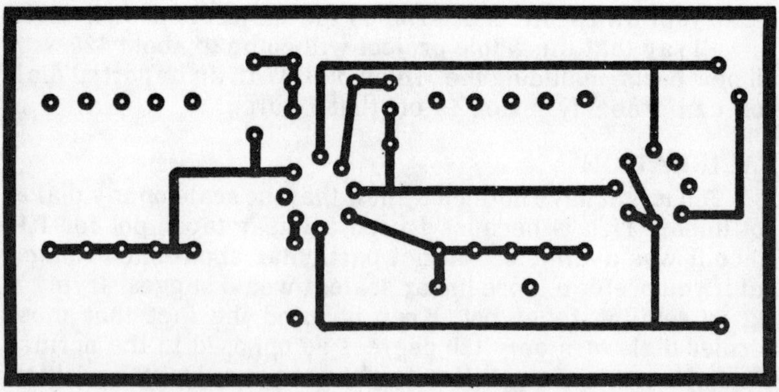

Fig. 6. Full-sized printed-circuit board used in the IC square-wave generator.

capacitors from my junkbox. Even so, they came out very close.

OPERATION

Square waves are very handy for testing amplifiers of all sorts in conjunction with an oscilloscope. In audio work, they will reveal poor high- or low-frequency response, ringing, and other ailments. Of course, you don't need a scope just for general testing of audio amplifiers and such. A simple signal tracer will do.

Speaking of oscilloscopes, you can use 500 kHz square waves for adjusting compensating capacitors in scope probes and step, or decade, attenuators. Usually, the instruction manual of your scope will outline the proper procedure. Since this unit will supply a signal at up to 1 MHz in frequency, it can be used to fix or test broadcast receivers. However, a detailed discussion of testing with square waves is beyond the scope of this article.

I'm sure that if you build this square-wave generator, you'll be very pleased with its performance and reliability.

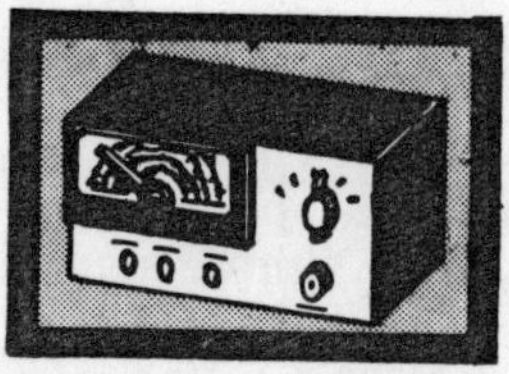

Pulse-Train Generator

Hank Olson

First let's find out what pulses are, and then find what one can do with them. A **pulse** (as the word applied to radio) is an electromagnetic wave or modulation thereof, of brief duration. This definition allows us a great deal of latitude. However, when most electronics-oriented people see the word "pulse," they think of a **rectangular** wave whose "on" time is short compared to its period.

SIMPLE PULSE GENERATORS

A pulse train is shown in Fig. 1, where t_o is the on time, and t_p is the period, or distance between similar points on adjacent pulses. A pulse is not required, in general, to be rectangular; it may have any of a variety of shapes, as shown in Fig. 2. These special-purpose pulses are sometimes used in pulse systems where it is desired to carefully control the bandwidth of the pulse signal. A pulse generator that will produce all of the pulses of Fig. 2 is not within the scope of this article; so we'll stick to the common rectangular pulses of Fig. 1.

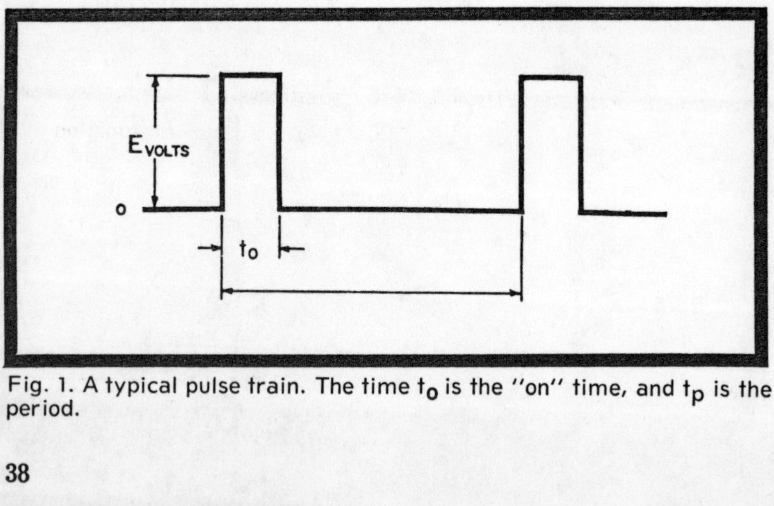

Fig. 1. A typical pulse train. The time t_o is the "on" time, and t_p is the period.

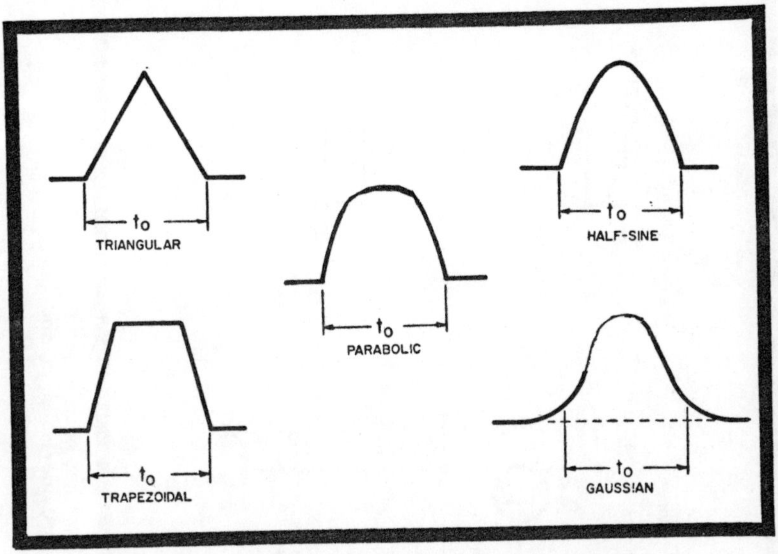

Fig. 2. Other pulse shapes that may be encountered.

Perhaps the simplest pulse generator known to solid-state electronics is the unijunction relaxation oscillator, shown in Fig. 3. It puts out a train of pulses whose repetition rate $(1/t_p)$ is controlled by R1 and C1, and whose on time (t_o) is controlled by R2 and C1. With a few slight modifications, the unijunction relaxation oscillator can be synchronized and the RC charging of the emitter capacitor modified to linear charging.

Since the circuit of Fig. 3 produces a pulse of rather short duration, we can use it to trigger a monostable multivibrator to obtain longer pulses. It is in such service that the monostable multivibrator is called a **pulse stretcher**. A realization of this type of pulse generator is shown in Fig. 4. A

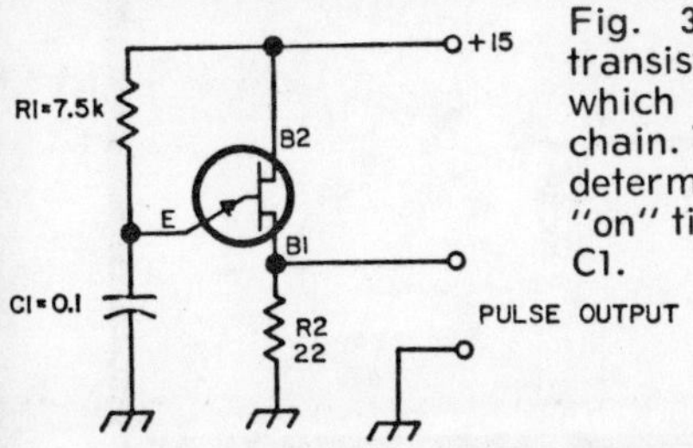

Fig. 3. A simple unijunction transistor relaxation oscillator, which will generate the pulse chain. The pulse repetition rate is determined by R1 and C1; the "on" time is controlled by R2 and C1.

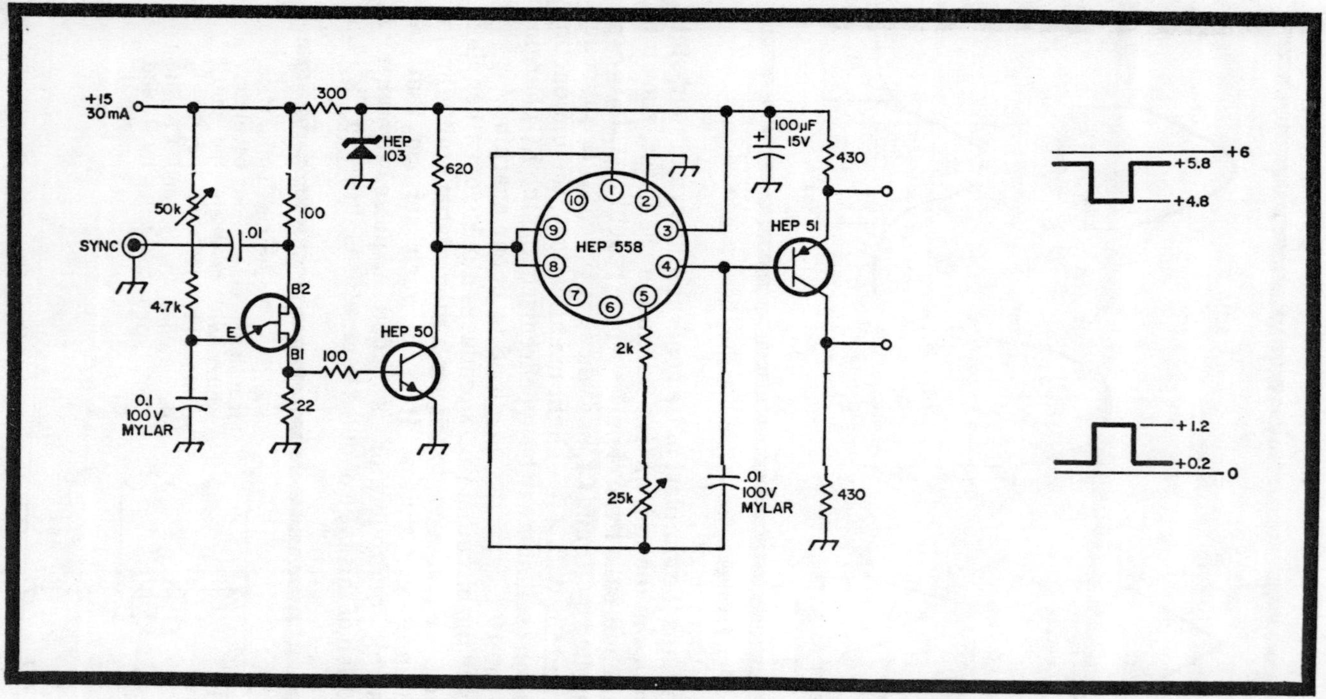

Fig. 4. In this pulse generator, a pulse amplifier has been added to the simple unijunction relaxation oscillator to provide isolation and the necessary shift level.

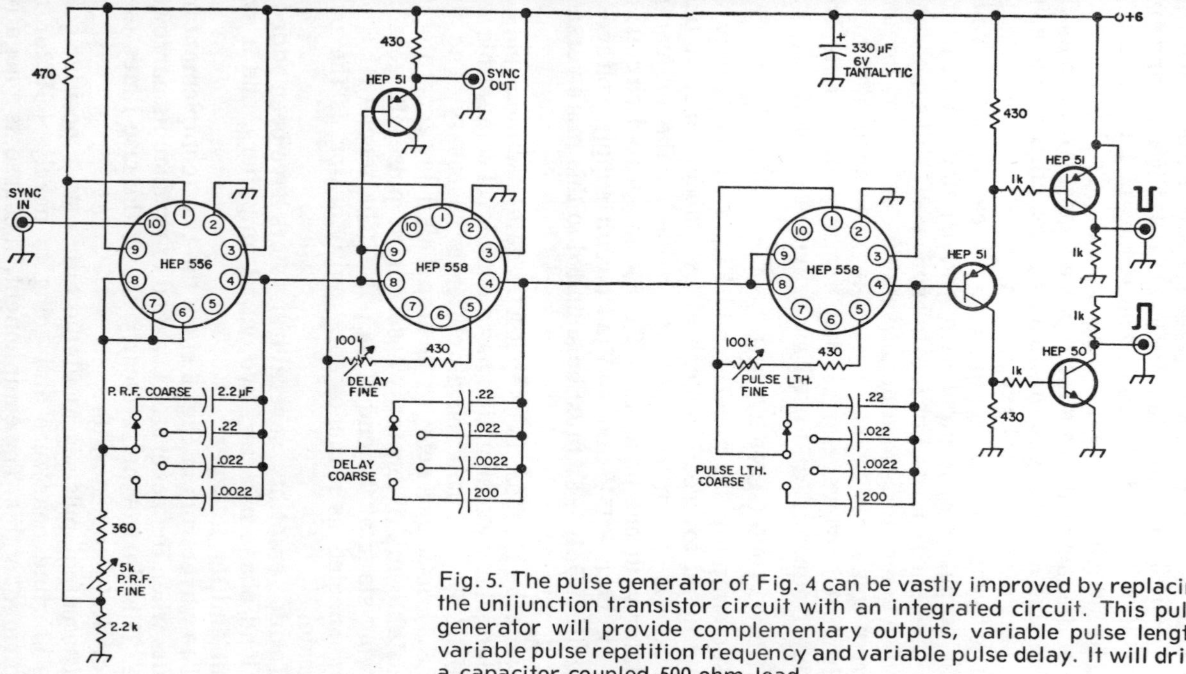

Fig. 5. The pulse generator of Fig. 4 can be vastly improved by replacing the unijunction transistor circuit with an integrated circuit. This pulse generator will provide complementary outputs, variable pulse length, variable pulse repetition frequency and variable pulse delay. It will drive a capacitor-coupled 500-ohm load.

pulse amplifier was added between the unijunction section and the monostable multivibrator to provide both isolation and the level shift necessary. The values shown in Fig. 4 will give pulse repetition frequencies (PRFs) of 100 to 1000 Hz and pulse widths of 30 usec to 500 usec. The values of R1C1 and R2C2 could be switched to provide other PRFs and pulse lengths. In spite of the fact that this simple pulse generator has only five semiconductor packages and about a dozen other components, it is a very useful unit to have around.

The "sync" connection to the circuit can be either used as a sync output or sync input. A negative-going wave, from a low-impedance source, put **into** the sync port, will synchronize the pulse generator. Or, a negative-going pulse may be taken out from this same "sync" port. The pulse generator can be synchronized by waveforms that are multiples of its free-running frequency, and used as a divide-by-n unit.

GENERAL USE GENERATOR

The simple pulse generator of Fig. 4 can be vastly improved upon, to create a general use model. Such a pulse generator is shown in Fig. 5. By replacing the unijunction oscillator with one using an HEP 556 integrated circuit, the need for both +15V and +6V as circuit supply voltages is eliminated. Only +6V must be supplied to this more versatile pulse generator.

Since the HEP 556 (a 3-input ECL gate) oscillator puts out a rectangular waveform that has a logic level compatible with the HEP 558, no isolation amplifier is necessary. The first HEP 558 is used as a delay generator. This delay generator is a monostable multivibrator whose output pulse triggers the following stage at the end of its pulse. The second HEP 558, also connected as a monostable multivibrator, is the pulse generator.

To add versatility, several transistors have been added: a split-load phase inverter, two output amplifiers, and a sync output emitter follower.

The waveforms in Fig. 6 show the two (complementary) outputs when the generator is asked to produce its narrowest pulse. This narrow pulse clearly shows the rise times to be expected of our generator.

The output pulse is available either as a positive-going pulse starting near zero and going to nearly +6V, or the complement of this. The complement, of course, is a negative-going pulse starting near +6V and going to nearly zero.

By switching the **pulse length** switch to **SQ**, the monostable multivibrator, which forms the pulses, becomes a simple

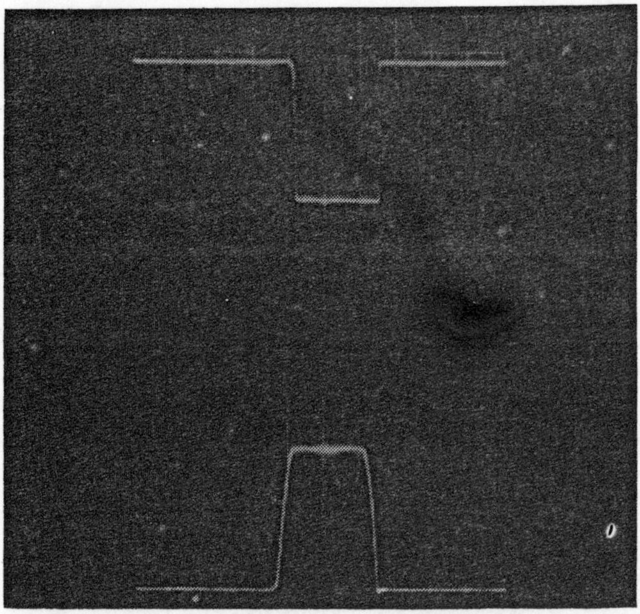

Fig. 6. Output waveform of the circuit in Fig. 5 for minimum pulse width. The sweep speed of the oscilloscope is 1 usec per division.

Fig. 7. The pulse generator. This unit uses integrated circuits and transistors to provide pulse repetition rates from 20 Hz to 200 kHz, pulse delays from 20 usec to 20 msec, and pulse widths from 20 usec to 20 msec. A square-wave output is also available.

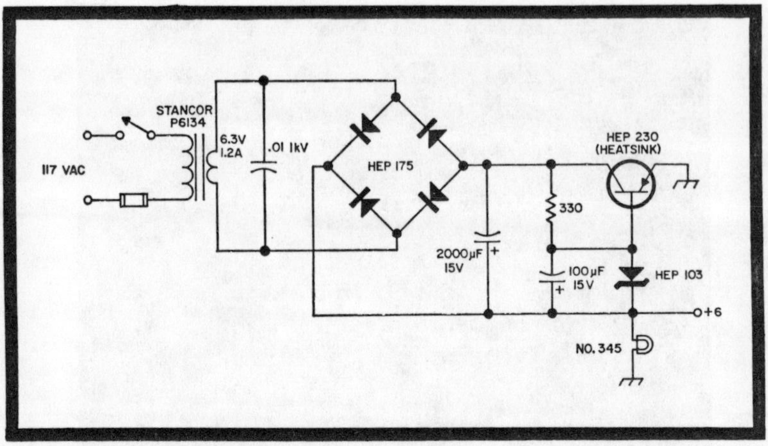

A regulated power supply for the pulse generator shown in Fig. 5.

divide-by-2 flip-flop. In this mode, nearly perfect square waves are produced from 10 Hz to 100 kHz—just half the normal PRF rate. This feature was added because it was so simple. The monostable multivibrators that perform the delay and pulse length functions are basically flip-flops modified for monostable use.

In the interest of simplicity, the four ports of the pulse generator are direct coupled. Coupling capacitors, of proper size to accommodate the particular pulse one is using, can be added externally. Alternately, a series dry cell or battery can be used to adjust levels—these are best used externally.

The finished pulse generator is shown in Figs. 7 and 8. The generator was built in a Bud cabinet (CD-1480) for two reasons. First, this was the cabinet the author had on hand; and second, the 8 x 8 in. panel allows enough room to mount all the controls. The circuit board picture, Fig. 8, shows all the generator circuitry except the power supply. Obviously, there is room to spare, and a much smaller unit could be built if miniature switches and pots were used.

APPLICATIONS

Now that we've generated our pulses, let's have a look at their uses. The most commonly used pulse is the square wave, which the second pulse generator will produce. Square waves are widely used to performance check audio amplifiers. A thorough section on square-wave testing of amplifiers is given in TAB book No. 702, "Electronic Measurements Simplified," by C. L. Hallmark.

Fig. 8. Circuit board used in the pulse generator shown in Fig. 7. All components are mounted on this board except the power supply.

Since the radio telegraph, radioteletype, and television modes of transmission are all based on pulse sequences, a pulse generator can be useful in the design, simulation, and testing of equipment for these modes. The exact nature of the use of the pulse will depend on what the user is trying to do. If he wished to turn on a transmitter with the pulse, we'd call it **pulse** (amplitude) **modulation**. If he wished to turn a signal off with a pulse, we'd call it **blanking**. Turning a signal on for a desired time interval, after a desired interval, would involve **delaying** and **gating**. Of course, the uses of combinations of these functions are limited only by the imagination.

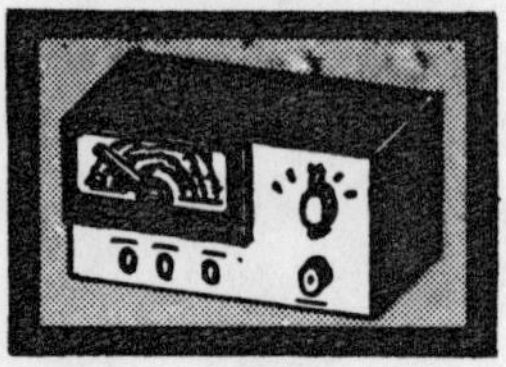

An IC Pulser

Hank Olson

In designing a pulse generator, the first consideration has to be compatibility with ICs. From the standpoint of convenience, economy, or versatility, it has become unattractive to build **anything** digital without IC logic. What this means to pulser design is that the pulser should be compatible with the types of digital ICs one finds himself using most.

Although RTL is the IC logic family most widely used by experimenters, it is extremely doubtful that this situation will last. The two types of current-sinking logic—DTL and TTL—are far out in front of RTL in industry usage and gaining daily. The reasons for industry preference of DTL and TTL over RTL are several: better noise immunity, higher speed, larger fanout, and a larger selection of devices around which to design.

PULSER THEORY

This pulser was designed so that its output would be compatible with DTL or TTL, and also with ECL. The implementation of these two outputs is simplified by use of ICs that are designed for buffer and clock driver service. (The pulse generator section is shown in Fig. 1, and the output section in Fig. 2.)

The free-running multivibrator that determines the basic "rep" rate, the delay one-shot, and the pulse length one-shot use HEP versions of MECL ICs.

By using an etched circuit board, it is possible to reduce lead lengths and get shorter pulses. Longer pulses are also provided by extending the range at the other end of the range switches. An additional transistor has been added to both the **delay** and the **pulse length** one-shots. These transistors allow the one-shots to recover more quickly, providing more stable operation for pulse lengths approaching the period of the basic rep rate.

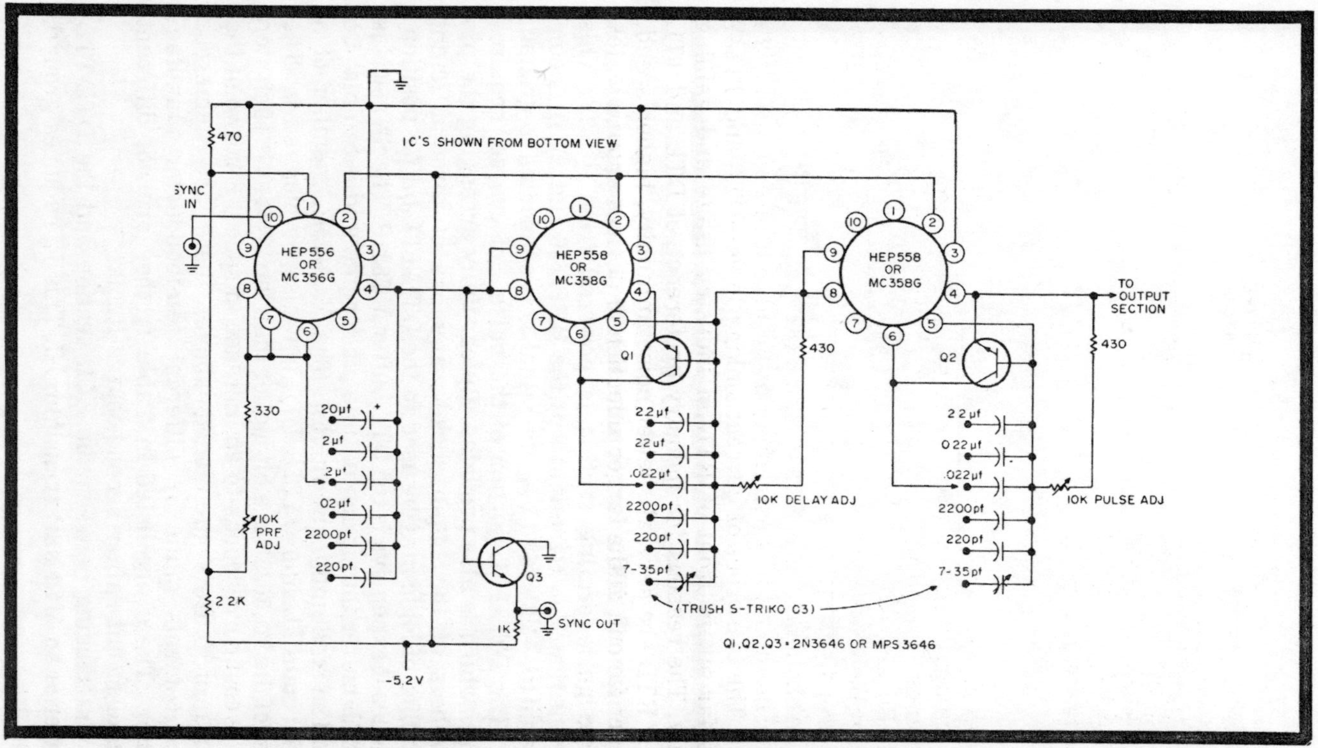

Fig. 1. Pulse generator portion of IC pulser.

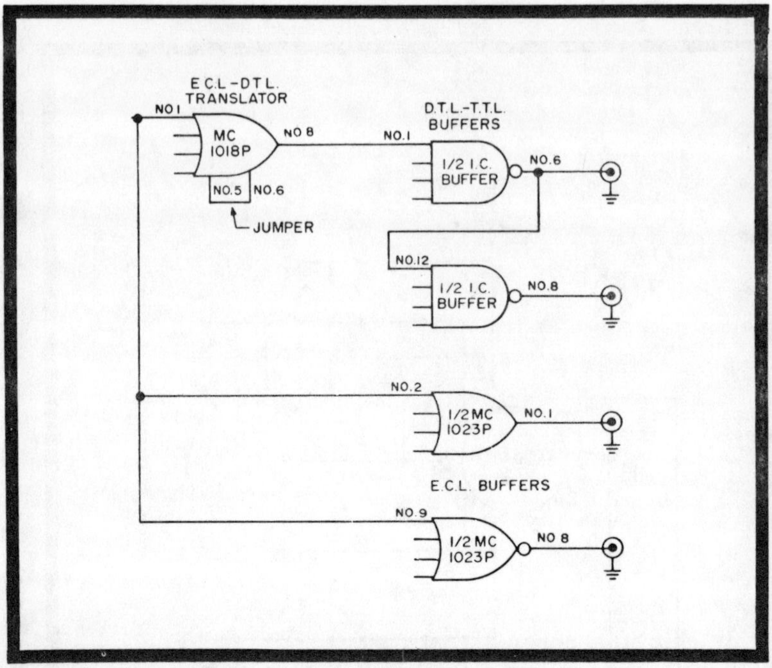

Fig. 2. Output section of pulser.

Since one of the two outputs is to be ECL compatible, the basic pulse-forming section is powered from —5.2V. This means that V_{ee} (terminal 2 of the HEP 556 and 558s) is connected to —5.2V, and V_{cc} (terminal 3) is connected to ground.

The DTL-TTL portion of the pulser, of course, requires +5V, which is also provided. By carefully arranging the pin numbers of the 14-pin socket, any one of three different families of logic may be used in the DTL-TTL buffer position. The least expensive is the DTL buffer MC832P. The SN7440N is a TTL buffer that is compatible pin-for-pin, and it can also be used at a slightly higher cost. The MC3025P, member of a different TTL line (MTTL III), can also be plugged in. The cost of the MC3025P is a bit higher than the SN7440N; each step up in cost corresponds to an increase in speed. For two of the types of buffers mentioned above, there are numerous replacements made by different semiconductor manufacturers. These are listed in Table I; they are not different families, but rather second-source items.

Interfacing between the ECL section and the DTL-TTL buffer is an MC1018P translator IC. This IC requires ground,

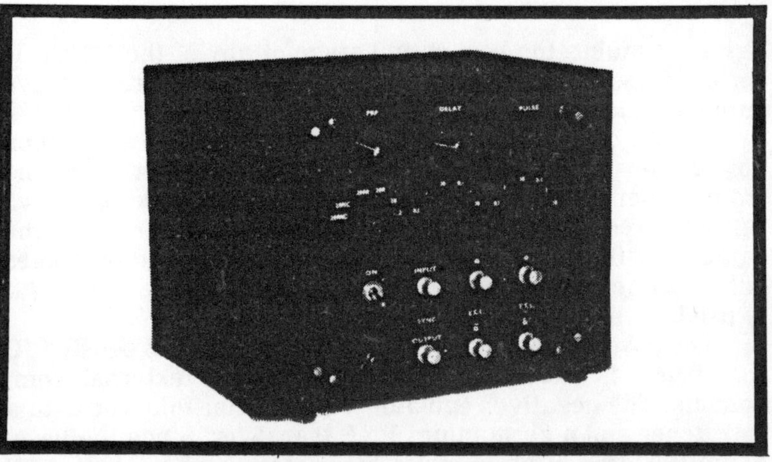

+5V, and —5.2V for supply connections. A simplified circuit of the MC1018P is shown in Fig. 3 as used in the pulser. Of course, such a circuit could be built of discrete components, but not as simply and inexpensively as by using the MC1018P. If the MC1018P is hard to find, the MC1018L may be used. It is the same chip in a **ceramic** case, and is being sold at a slightly higher price than the plastic unit.

The output section for the ECL compatible pulses is provided by an MC1023P. Although this IC is billed as a **clock**

Table I. IC Sources

Specified Type	Manufacturer	Equivalent	
SN7440N	Texas Inst.	DM 8040N	National Semiconductor
SN7440N	Texas Inst.	USN 7440A	Sprague
SN7440N	Texas Inst.	N 7440A	Signetics
SN7440N	Texas Inst.	FJH 141	Amperex
SN7440N	Texas Inst.	MC7440P	Motorola
MC832P	Motorola	DT uL 932	Fairchild
MC832P	Motorola	SW932-2	Stewart Warner
MC832P	Motorola	DTL 932	Sperry
MC832P	Motorola	PD 9932	Philco
MC832P	Motorola	SN15 832N	Texas Inst.
MC832P	Motorola	S 9323	Sylvania
MC832P	Motorola	MIC 932	ITT
MC832P	Motorola	HSC 932	Hughes
MC832P	Motorola	CD2306E / 832	RCA
MC 832P	Motorola	RM 932	Raytheon

driver, it makes the best output stage of any of the MECL II series. This is true because of its exceptional ability to drive capacitive loads.

The regulated +5V and —5.2V are provided by a common power transformer and rectifier. Since the centertap of the transformer is grounded, the circuit may be considered as two full-wave rectifiers (one positive and one negative) across the same transformer. Of course, the diode configuration comes out the same as a full-wave bridge; so an IC bridge (HEP 175) is used.

The positive regulator is an MC1460R, an inexpensive IC that provides excellent regulation with few external components. The negative regulator is an emitter follower with a 5.6V zener and a germanium PNP transistor. Since the base-to-emitter drop of germanium transistors is about 0.3V, 5.6V minus 0.3V gives us close to the 5.2V required for the ECL ICs.

Both the MC1460R and the HEP 232 (PNP power transistor) are diamond-shaped and meant to be fastened to a heatsink. They each have an aluminum bracket attached to them to fulfill this requirement for a dissipator.

The entire circuit of the pulser is built on an etched circuit board. In fact, the wafer switches are assembled so that the

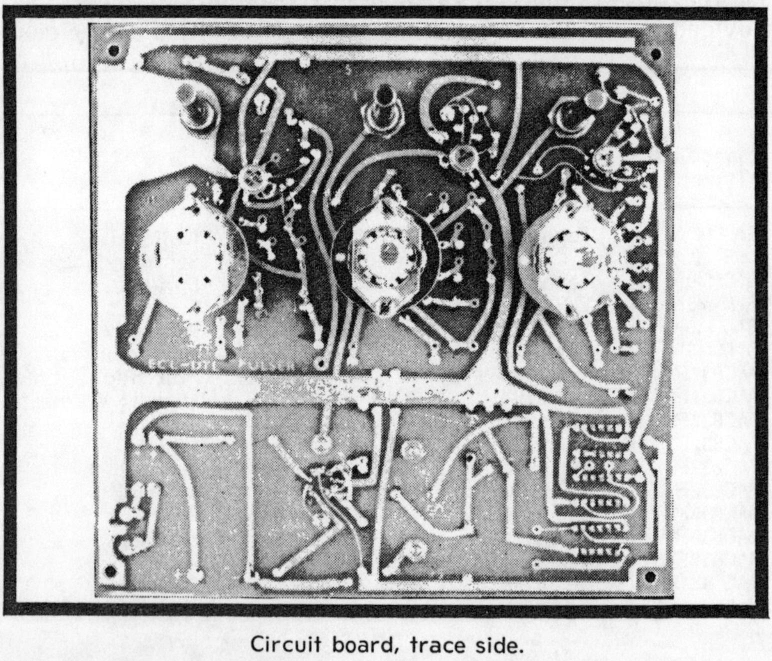

Circuit board, trace side.

Circuit board, parts side.

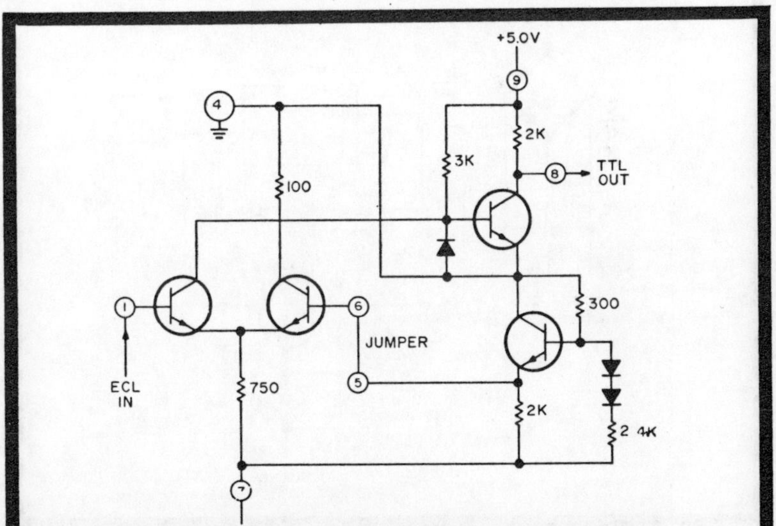

Fig. 3. Simplified ECL-TTL translator circuit.

board is clamped into the switch assembly. The shafts of the switches are cut to ⅞ in. to extend out the same length as the pot shafts. The entire board is then mounted to a panel using 1⅛ in. spacers. Some care must be exercised not to allow the spacers to short any of the traces of the circuit board to ground. This can be insured by using fiber washers between board and spacers. The spacer in the power supply corner is intended to connect the board ground to the panel, so no fiber washer should be used at that corner.

All the components are mounted on the side of the board away from the panel, except the three HEP ICs. These three ICs were reverse mounted to ease layout, allowing shorter trace lengths.

The 7⅞ x 7⅞ in. panel is made to fit a Bud CD-1480 cabinet, which has enough panel space for all the controls and jacks. The power transformer is mounted (off the board) inside this cabinet.

As to the DTL-TTL output capability of the pulser, it is dependent on the exact type of output stage. Each of the ECL outputs will drive 24 ECL gates (has an ECL fanout of 24). But the fanout of each of the two current-sinking logic outputs is as follows: MC832P, 24 DTL load units; SN7440N, 29 TTL (SN7400N series) load units; MC3025P, 19 MTTL III load units.

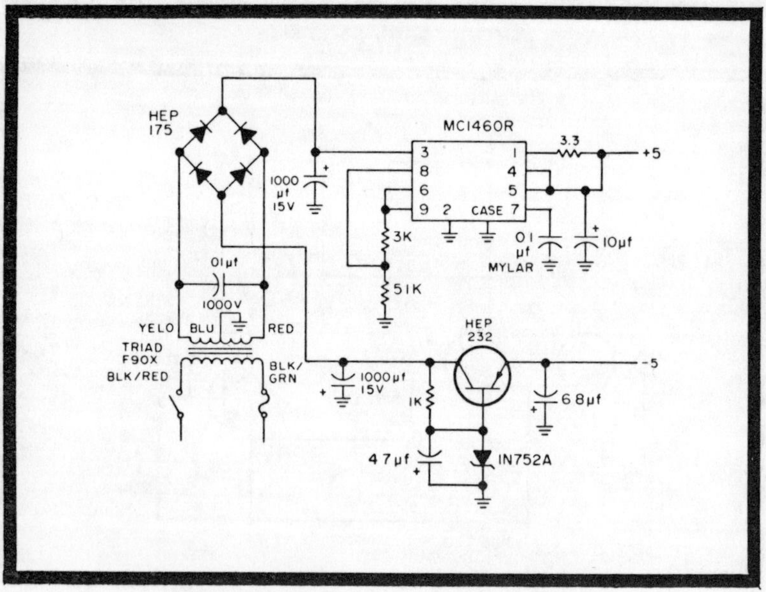

Fig. 4. Power supply for ECL pulser.

Fig. 5. PC layout of IC pulser.

The load units are not the same for these three current-sinking families, so it is best to use the type of output IC for the sort of family **you** use most. It is quite all right to use any of the types of current-sinking logic ICs with the pulser (no matter which IC is used in the output stage), but some reduction of fanout may be experienced with certain combinations.

PERFORMANCE

The pulser will produce pulses (and delays) from about 50 nsec to 30 msec. The **delay** and **pulse** switches each have six positions, and the variable control associated with each continuously varies each over at least a ten-to-one range. Pulse repetition rate is adjustable from about 0.5 Hz to 1 MHz with a 6 position switch and variable control. The labeling of the three panel switches was minimal, with 200 kHz to 2 Hz (in that order, so that the period steps in the same direction as delay and pulse width) for the rep rate. These labeled frequencies correspond to frequencies within each switch position, not the center or either end point. The delay and pulse switch positions were similarly labeled from 0.1 usec to 10 msec for simplicity.

CONSTRUCTION

The entire pulser is built on an etched circuit board, including the power supply, whose circuit appears in Fig. 4. Figure 5 is a half-scale copy of the board.

Since the etched-board method has been used, some specialization in components is necessary. The Cornell-Dubilier BR1000-15 filter capacitors are physically smaller than most other brands of the same capacity and voltage

Fig. 6. Component placement (etched side of board shown).

rating, and so are best used. Similarly, the switches are Centralab PA-1 and PA-300 combinations, and the two variable capacitors are Trush S-Triko 03 types. The IC sockets were HEP 451 for the round can types and Methode M1141 for the dual-inline types. Figure 6 is the layout of all components on the board.

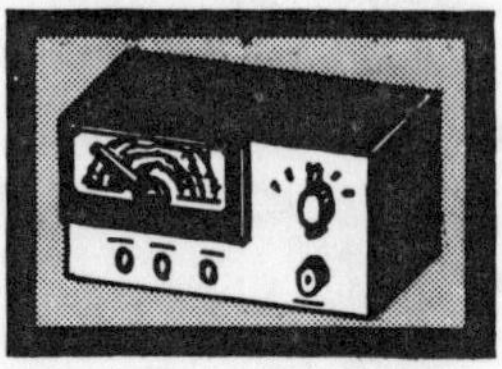

100 kHz Marker Generator

W. W. Davey

This useful piece of equipment generates usable harmonics from 100 kHz to 225 MHz. It is completely self-contained and portable, which makes it convenient not only to use in the shop, but also in mobile units. Its use lies mainly in accurately spotting band edges and 100 kHz calibration points. The circuit diagram is shown in Fig. 1.

Most modern-day communications receivers are equipped with calibrators, but these calibrators are of little use for spotting frequencies on VHF and UHF converters or portable equipment.

The generator is constructed in a 2¼ x 4¼ x 1½ in. box. The parts are mounted on a Vectorbord, and the entire unit is powered by one No. 216 9V battery or its equivalent.

HINTS ON CONSTRUCTION

First obtain some Vectorboard. The piece I used was cut from the board supplied in a GE experimenter and hobbyist

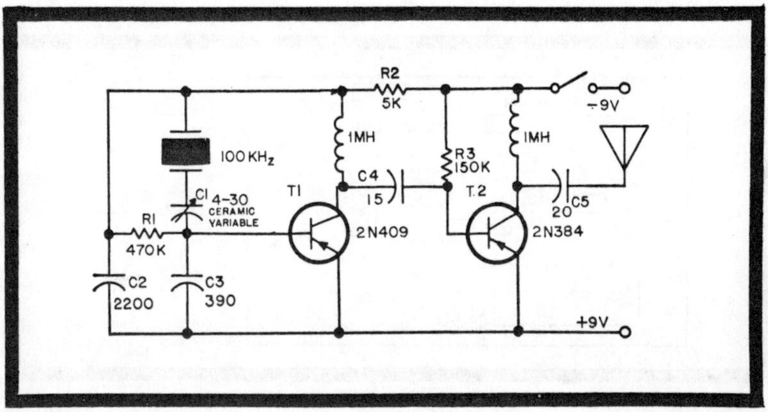

Fig. 1. Circuit diagram.

kit. The board must be cut to size before construction and will measure 3½ by 2 in. This will allow room for the 9V battery in the end of the box. Make sure the newly cut Vectorbord will fit inside the box before you start mounting parts. It might save a lot of trimming at a later date.

The parts layout is not critical. Components may be arranged as shown in the photo (and Fig. 2) or in any other arrangement suitable to the components you may be using. I used sockets for the transistors, as I wanted to be able to experiment to see which transistors from my junk box would give the most output in the VHF and UHF bands. I ended up using the 2N404 for the oscillator and a 2N384 in the multiplier stage. I also found that Japanese 2SA83 transistors removed from the i-f stages of a junked transistor radio, would work equally well in both sockets.

All components are mounted on the top of the Vectorboard, with the exception of C4 and C5. For the most part, wiring can be completed with existing leads on components. The push-in terminals furnished with the GE experimenter kit were used for the battery and antenna connections, and for mounting the crystal socket. The circuit board can be mounted to the box with three 11 / 16 in. bushings. This leaves room for a slide switch to be mounted on the cover. Two of these bushings were purposely placed at the end of the board to form a sort of socket to hold the 9V battery. The antenna output connector mounted on the box is a Switchcraft No. 3501FP phono jack. A small hole may be drilled in the bottom of the box and a screwdriver may be inserted for adjusting C1. For extreme accuracy, C1 is adjusted to zero beat with WWV.

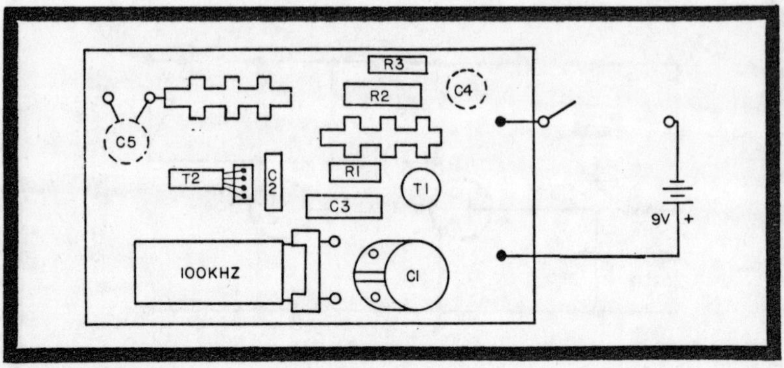

Fig. 2. Parts layout.

A 36 in. piece of insulated wire soldered into a phono plug may be inserted into the phono jack and used as a test antenna. The intensity of the markers may be varied by moving this test antenna near your receiver antenna lead-in. If you are using a coax lead-in, you can couple by drilling a small hole in your coax relay so that the test antenna can be inserted near the relay armature. You will find that most swr meters provide an easy method of coupling to the center conductor of the coax. As a last resort, you can always couple to the receiver or converter antenna coil.

I-F Sweep Frequency Generator

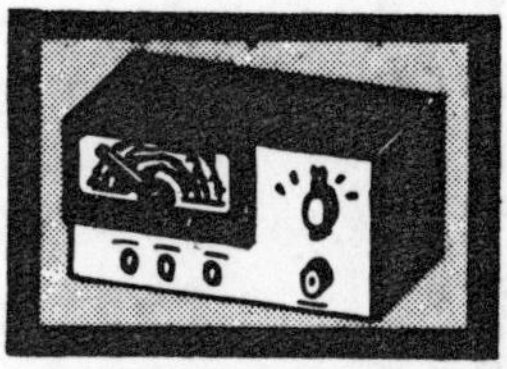

Edward A. Lawrence

Here is a simple sweep frequency generator for aligning the most common i-f strips. The unit has only one transistor because the sweep voltage is taken from the oscilloscope it is used with. Most general-purpose oscilloscopes have a sawtooth output jack. By using this sawtooth, the frequency is locked to the position of the trace.

OPERATION

The oscillator is made to deviate in frequency in step with the voltage applied to the base bias circuit, which is either the

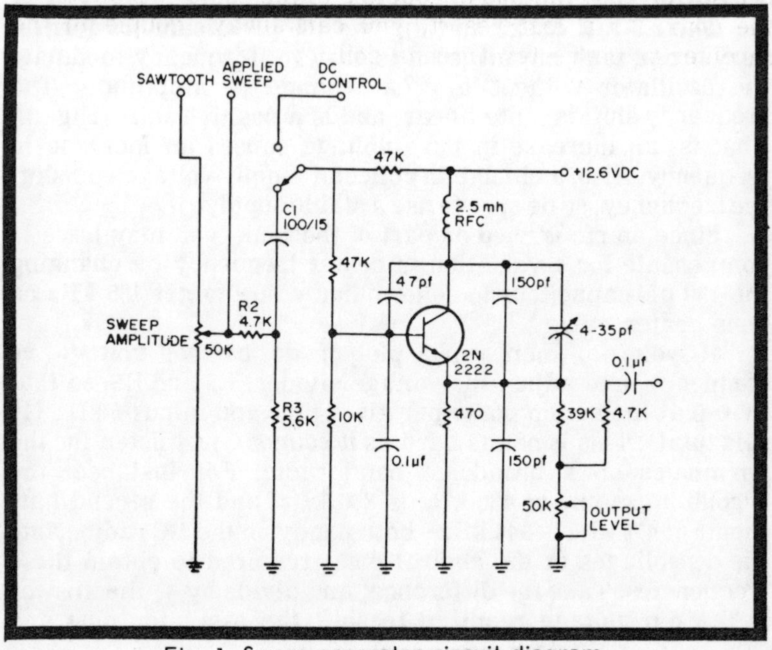

Fig. 1. Sweep generator circuit diagram.

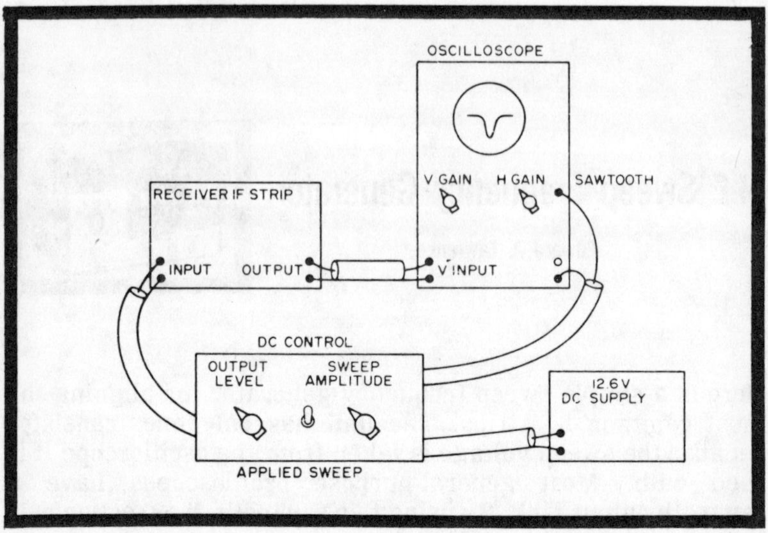

Fig. 2. Interconnections of sweep generator.

sweep voltage or the dc voltage applied at the control point. If no sweep or control voltage is applied, the oscillator runs at the center frequency and may be used as a conventional signal generator.

The 2.5 mH choke and the two 150 pF capacitors form a very broad tank circuit, so it is easy to frequency modulate the oscillator without a great change in amplitude. The frequency shift is quite linear, and is a positive shift (Fig. 2). That is, an increase in base voltage causes an increase in frequency. Also, a change in collector supply voltage will shift the frequency, so be sure to use a stable supply.

Since an rfc is used as part of the tank, you may have to compensate for a variation in center frequency by changing the 150 pF capacitors to some other value to get 455 kHz as your center value.

If you do, then run a plot of dc control voltage vs frequency, and adjust the voltage divider, R2 and R3, so that 1V p-p at the sweep pot wiper gives plus and minus 5 kHz (10 kHz total). This is not as hard as it sounds...just listen for the harmonics on a broadcast band radio. For instance, the second harmonic of 430 kHz is 860 kHz, and the second harmonic of 470 kHz is 940 kHz—both handy on the BC radio. Note the dc voltages at the control point required to obtain these frequencies. Take the difference, and divide by 4; the answer is the p-p voltage required to shift the oscillator plus and minus 5 kHz. Then apply a low, known ac voltage to the wiper of the sweep pot with the switch in the **sweep applied** position. Adjust the value of the divider resistors to get the proper fraction. In my case it was 0.43.

It is advisable to run the sweep rate as slowly as possible, in order to display the response curve as accurately as possible. The sharper the skirts, the more slowly the generator must pass through the bandpass. With this generator, the amount of frequency deviation is controlled by the p-p amplitude at the sweep pot arm. If you were looking at the response of a regular i-f, you would set the sweep amplitude high to see the entire response curve. As far as an accurate display is concerned, this is fine, since the slope of the skirts is shallow. But if you were looking at the response of a sharp filter, you could not tolerate such a wide sweep, because the fast rate of change would tend to skew the display. To correct this, reduce the sweep amplitude to reduce the frequency deviation down to the edges of the skirts of the response curve. This reduces the rate of change and minimizes the skewing of the display. Also, it is better to display the i-f before detection, if possible, to prevent the detector time constant from possibly distorting the display.

If you have the sweep generator set up as described, it is easy to set the sweep for a known deviation and then proceed to read the 3 or 6 dB points from the face of the scope. Be sure to disable the agc for this test.

For those who aren't familiar with the setup for obtaining the response curve display, refer to Fig. 3 and the following outline.

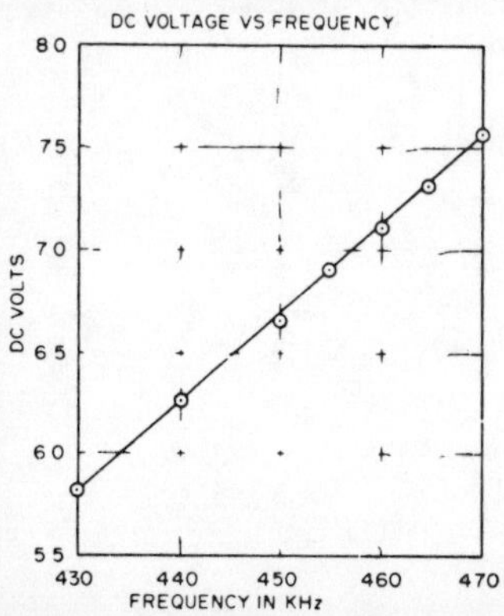

DC CONTROL VOLTS	FREQUENCY
5.86	430
6.27	440
6.69	450
6.90	455
7.12	460
7.33	465
7.55	470

Fig. 3. Sweep frequency vs dc voltage.

OBTAINING A RESPONSE CURVE

First, hook up the equipment as in Fig. 1. Set the **horiz gain** control for desired sweep width. Adjust the **sweep amplitude** pot on generator for desired frequency range. Set the **output level** to midrange. Adjust the **vert gain** control for the desired pattern height.

If you wish to change the total frequency deviation and the horizontal display width at the same time, use the scope's **horiz gain** control. This assumes that the amplitude of the sawtooth output is also varied by the **horiz gain** control.

If you wish to change the frequency spread and not change the width of the scope display, adjust the **sweep amplitude** control. This allows you to take a better look at the side lobes or any ripples in the passband, depending on how the controls are adjusted.

You don't have to use a sawtooth to sweep with if it isn't handy.

You can also use 60 Hz, but it will probably skew the passband you are trying to display, so I don't recommend it. One note here: When the sweep generator is first turned on, or the setting of the sweep amplitude is changed, the frequency will drift for a few seconds. This is due to the charge on C1 changing to a new level. Capacitor C1 is large to couple the low-frequency sweep with as little distortion as possible.

Wideband Frequency Generator

P.A. Lovelock

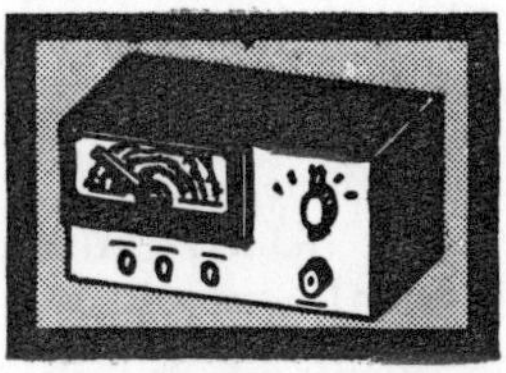

Ever been frustrated by a pyramid of test gear and an octopus of tangled leads and power cords while trying to align or troubleshoot your receiver? Or perhaps, the minimal space of modern apartment dwelling has prevented you from harboring much in the way of test equipment. In either event, you can use a "Gentrac."

Originally conceived as a "do all" test set for receiver work, the Gentrac also possesses useful capabilities for transmitter and audio circuit testing in a single, transistorized package that can be used anywhere.

SPECIFICATIONS

The Gentrac incorporates two individual units from which its name is derived: (1) a multipurpose generator, furnishing

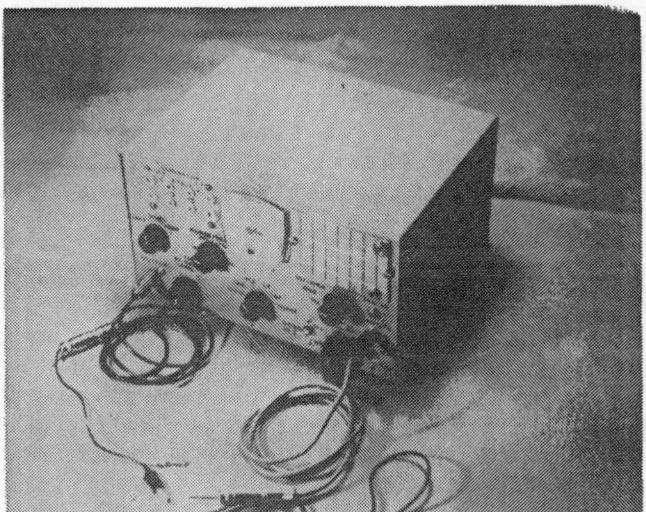

the most needed signals for communications work, and (2) an rf-af signal tracer and output meter.

Only two test leads are required, one for generator output and the other for tracer input—eliminating the usual cat's cradle that confounds the more conventional setup.

Here is what the Gentrac has to offer in terms of generator specifications:

1. 100 kHz and 1 MHz calibrated markers up to 200 MHz; crystal-controlled oscillator output in the range of 100 kHz to 30 MHz on fundamentals, up to 200 MHz on harmonics.

2. Audio sine and square waves over the voice frequency range of 250 to 2500 Hz.

3. Broadband "white" noise.

4. Modulation of rf output by 250 to 2500 Hz sine wave.

5. Generator output of low-impedance continuously variable over a 60 dB range for all signals.

The tracer specifications are as follows:

1. Input impedance: selectable 1M or 10K.

2. Input sensitivity (gain): selectable 150 uV or 3 mV.

3. Af input: Direct.

4. Output: Selectable to internal speaker or VU meter.

GENERATOR CIRCUIT DESCRIPTION

A schematic of the generator is shown in Fig. 1. For convenience, the circuit has been divided into six basic sections: G1, G2, G3, G4, G5, and G0. These individual oscillator and output circuits are as follows:

G1 Marker Generator. Stage Q1 is a dual-controlled oscillator, selectable for 100 kHz and 1 MHz output. Capacitors C2 and C3 permit accurate frequency adjustment. Stage Q2 is a harmonic amplifier for the 100 kHz output only, providing markers well up into the VHF range.

G2 HF Generator. Transistors Q3 and Q4 comprise an untuned crystal oscillator functioning over the range of 3 to 30 MHz.

G3 LF Generator. Transistor Q5 is an untuned crystal oscillator functioning over the range of 100 kHz to 4 MHz.

G4 Af Generator. Stage Q7 is an RC-coupled, phase-shift oscillator, with good quality sine-wave output, continuously variable over the 250 to 2500 Hz range. This oscillator also drives Q8 and Q9, which comprise a square-wave amplifier.

G5 Noise Generator. Diode D2 is a conventional noise generator, utilizing either a 1N21 or 1N23. The diode current regulator resistor, R6A, is a potentiometer section ganged to the output attenuator, R6B. Thus, while the output attenuator is not used with the noise generator, the noise output is adjustable with the same panel control. Voltage supply for the noise diode is stabilized by a 6V zener diode (D1), so that noise output is unaffected by normal battery variation and permits control setting to be used as a direct reference when making relative noise measurements. This eliminates the need for the usual diode current meter for reference purposes.

G0 Output-Attenuator Stage. Transistor Q10 is an emitter follower, which is the common output stage for G1, G2, G3, and G4. This stage provides low-impedance output with negligible loading on the individual generator circuits. The output attenuator, R6B, also functions as the emitter resistor for Q10. The generator mode switch, S1A, B, C, and D selects the appropriate generator-signal to be fed to the base of Q10, and also switches generator supply voltage. Sine-wave modulation of Q10 is provided by diode-modulator D5 when the **af** mode switch, S2, is switched to the **mod** position.

TRACER CIRCUIT DESCRIPTION

Figure 2 is a schematic of the tracer section, which is comprised of a commercially available transistor amplifier

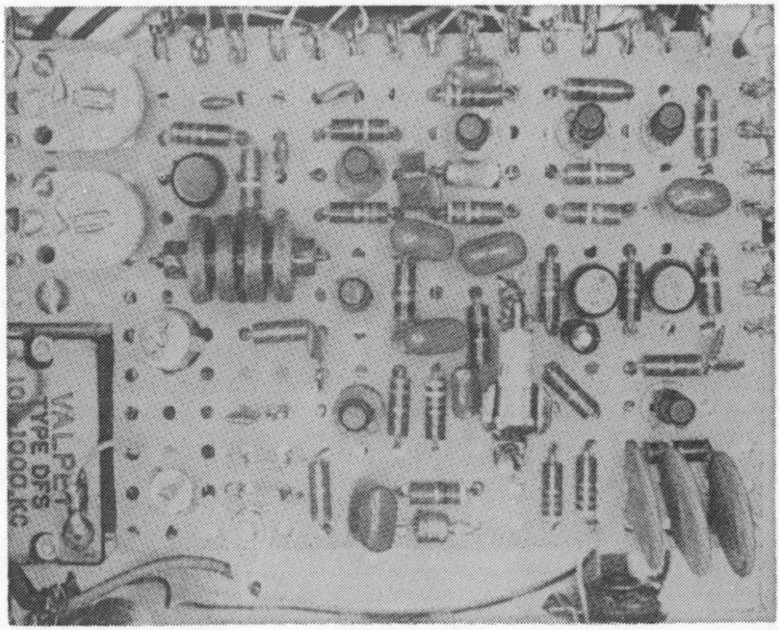

Generator section component board layout.

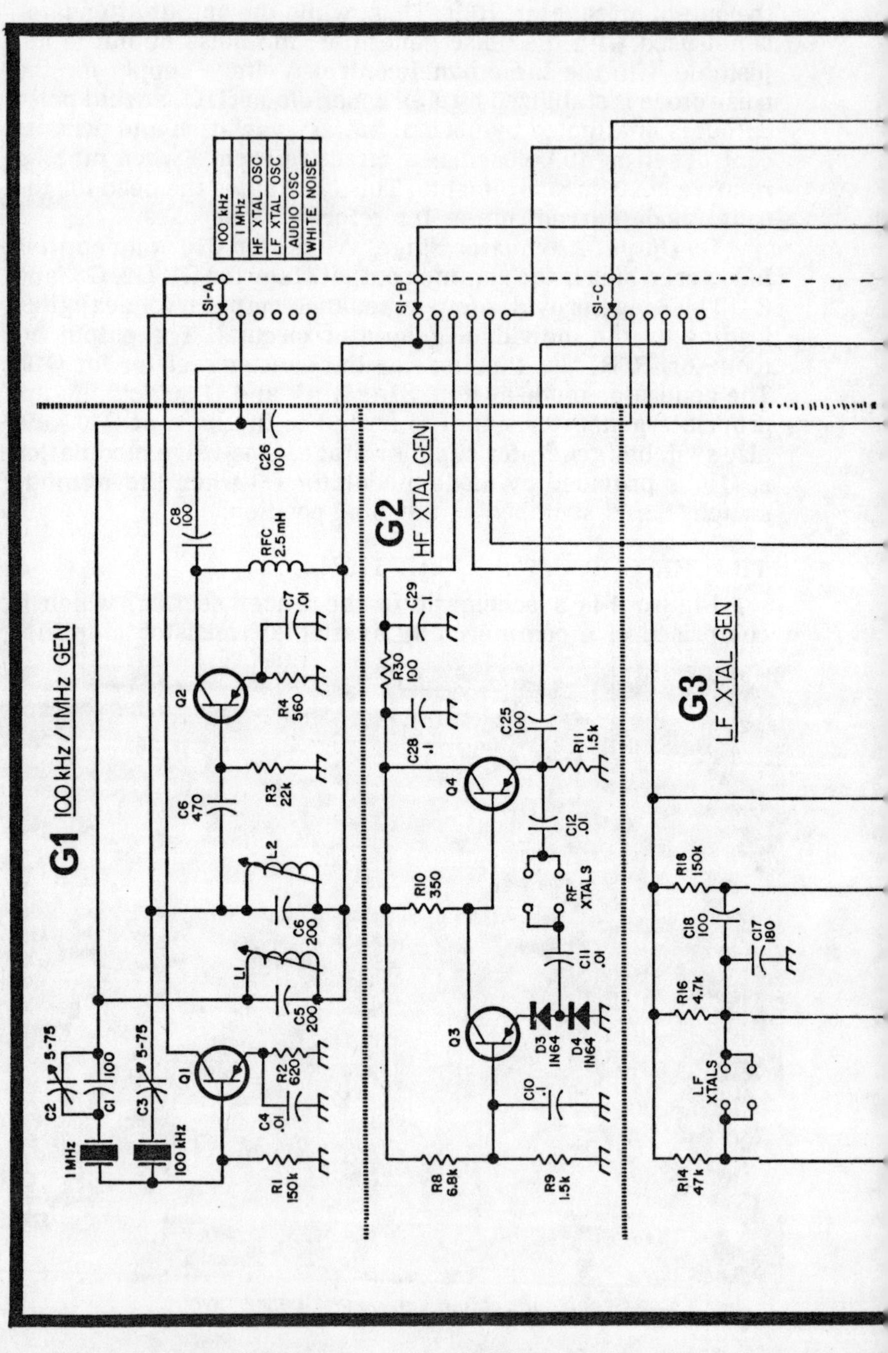
G1 100 kHz / 1MHz GEN
G2 HF XTAL GEN
G3 LF XTAL GEN
100 kHz
1 MHz
HF XTAL OSC
LF XTAL OSC
AUDIO OSC
WHITE NOISE
S1-A
S1-B
S1-C
C2 5-75
C1 100
C3 5-75
1 MHz
100 kHz
Q1
Q2
Q3
Q4
R1 150k
R2 620
C4 .01
C5 200
L1
C6 200
L2
C6 470
R3 22k
R4 560
C7 .01
RFC 2.5 mH
C8 100
C26 100
R8 6.8k
R9 1.5k
C10 .1
D3 1N64
D4 1N64
C11 .01
RF XTALS
R10 350
C12 .01
R11 1.5k
C25 100
C28 .1
R30 100
C29 .1
R14 47k
LF XTALS
R16 4.7k
C17 180
C18 100
R18 150k

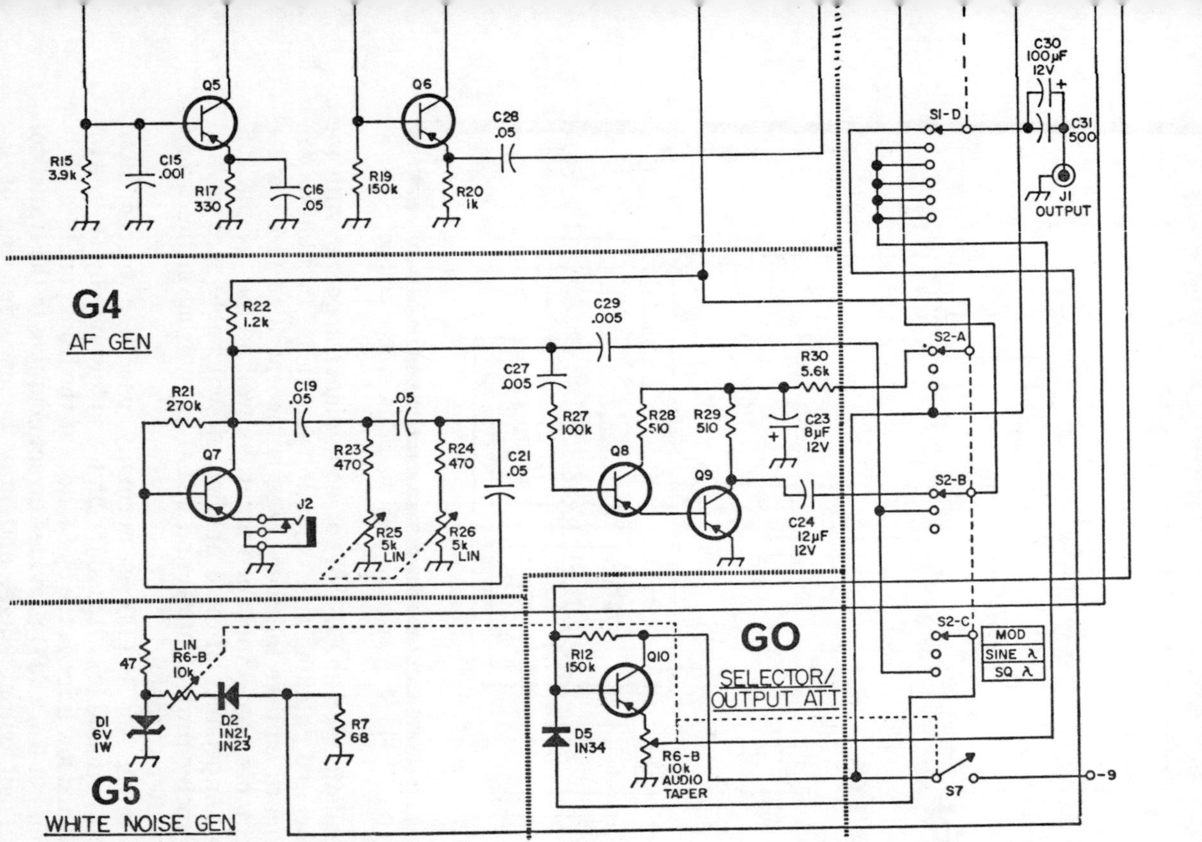

Fig. 1. The schematic of the generator.

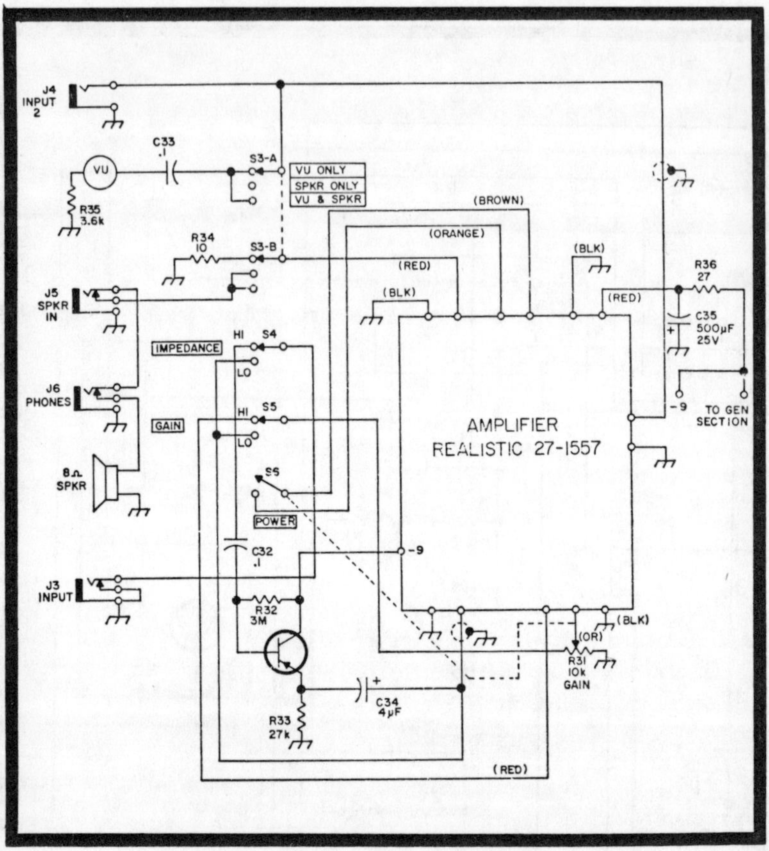

Fig. 2. Schematic diagram of the tracer section.

(Realistic type no. 27-1557, or equivalent), together with the associated patching and switching functions. In addition, an outboard emitter follower stage is incorporated to allow for selectable high-impedance input, as well as the normal 10K input impedance of the amplifier. A switchable speaker and VU meter provide aural and visual monitoring of the tracer amplifier output.

Input No. 1 (J3) is normally used for tracer input operation, and is switched for high and low gain by S5, and for high and low impedance by S4, both on the front panel.

Input No. 2 (J4) provides direct coupling to the transistor push-pull output stage, and may be used for one of the following purposes:

(1) A substitute output stage and speaker.

(2) Direct input to the VU meter (with tracer amplifier switched off) for measurement of high-level signals.

(3) A tracer amplifier output for high-impedance phones.

(4) Output for an oscilloscope to observe low-level rf or af signals picked up by the tracer.

Jack J5 permits using the internal speaker as a substitute with an external amplifier, and disconnects it from the tracer amplifier.

Jack J6 provides an output for low-impedance headphones, disconnecting the internal speaker.

Decoupling of the internal battery supply is accomplished by a 27-ohm resistor and a 500 uF capacitor, to prevent interaction between the generator and tracer when used simultaneously.

CONSTRUCTION

The entire generator circuitry, excluding front panel controls and crystal sockets, is mounted on a 5¼ x 3½ in. piece of perfboard (0.093 in. holes). Layout of the components is shown in the photo.

Good-grade, mica-filled sockets were used for transistors. Alternatively, transistor leads may be permanently soldered to flea clips staked in the perfboard.

The Gentrac, interior view.

The generator board and tracer amplifier are mounted in the 4½ x 8 x 6 in. aluminum mini box by means of standoffs. The standoffs are assembled from ½ and 1 in. spacers, normally intended for expandable 3-hole binders, and obtainable from any good stationary store. These spacers have 8-32 threaded studs at one end and threaded inserts at the other end, accepting 8-32 screws and nuts for attachment to the boards and box. The spacers may be connected together to assemble the desired length. Two 1 in. and one ½ in. length make up each of the four spacers to mount the generator board. Four ½ in. spacers support the tracer amplifier and two 1 in. spacers attach the input emitter follower to two of the tracer amplifier mountings.

The crystal used in the G1 marker generator is a Bliley dual 100 kHz-1 MHz unit, which was available in the author's junkbox. In the event difficulty is encountered in obtaining this unit, individual crystals may be mounted in the same board space. All the generator circuits were adjusted to permit common usage of the GE2 and GE9 transistors, since these are commonly available just about anywhere. However, other suitable transistors may be substituted, and the individual constructor may want to experiment with what is available. Substitutes for the GE2 may be any audio frequency PNP transistor with an hfe of about 70. Substitutes for the GE9 should be high-frequency PNP types, preferably good to 100 MHz, with an h_{fe} around 100.

Location of controls is shown in the front-panel photo. The HC6/U and FT243 crystal sockets are mounted on the front panel for both the HF and LF generator, in order to allow almost any crystal to be plugged in.

The Gentrac is powered by a Burgess D6, 9V battery, or equivalent. Smaller 9V batteries are not recommended for reliable operation, particularly since the drain of the noise generator is substantial. The battery is secured to the back of the minibox by a clamp fabricated from a wire clothes hanger. The springiness of this wire holds the battery securely, while permitting it to be removed when required.

The two test leads and rf test probe are assembled as shown in Fig. 3. Any small-diameter, shielded, single-conductor cable may be used for the tracer cable. For the generator output cable, miniature coaxial cable (type RG 174/U) is recommended. If this is hard to come by, good-quality, shielded single-conductor cable may be used, with some loss at higher frequencies.

The 1N34 diode used in the rf probe is of the larger glass type, since this fits snugly in the phono connector, with one glass seal forming an insulator between the wire-lead probe and the connector housing.

USING THE GENTRAC

The first question may be, "Why use crystal oscillators for the generator?" Obviously the Gentrac does not substitute for

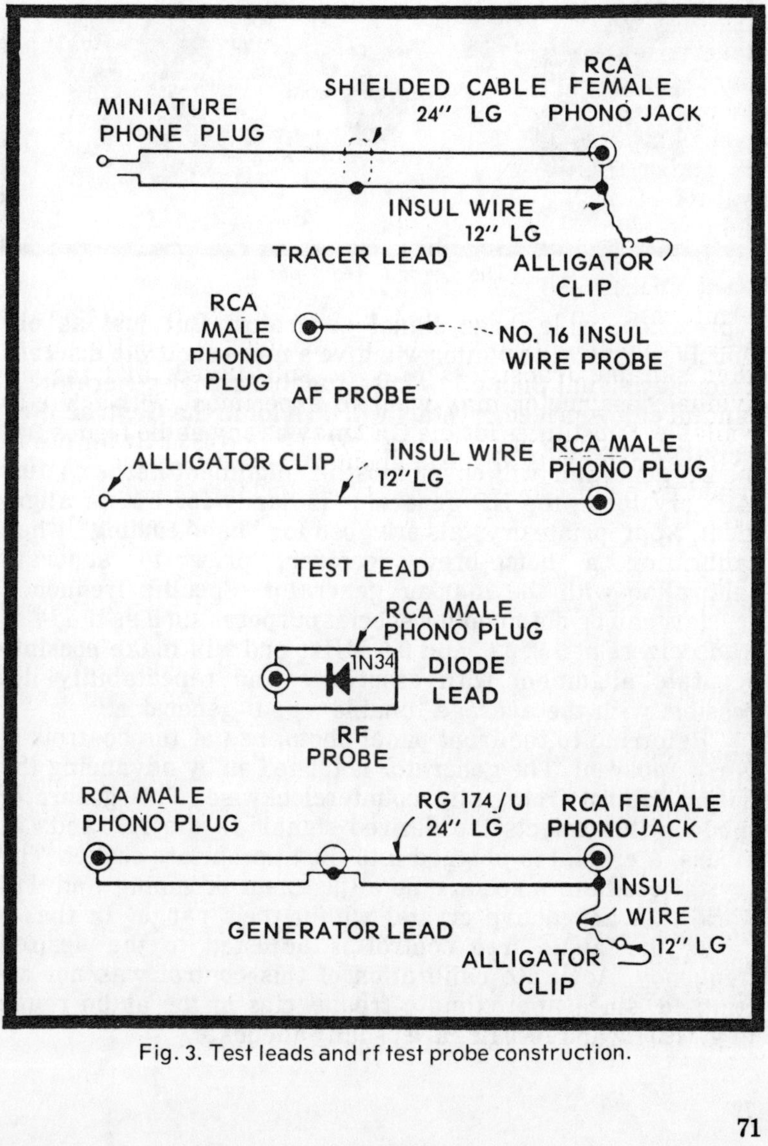

Fig. 3. Test leads and rf test probe construction.

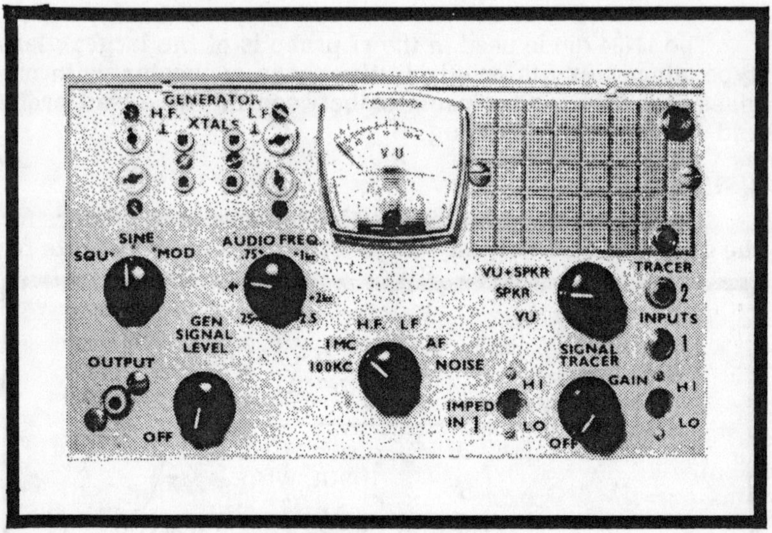

The Gentrac front panel.

a precision, wide-range signal generator. But just as obviously, a crystal oscillator will give a more accurate discrete frequency signal than a low-cost, tunable signal generator. A handful of low-priced crystals will do a lot in the Gentrac. For example, a couple of crystals at 262 kHz and 455 kHz for use in the LF generator will cover most i-f alignment needs. A 10.7 MHz crystal in the HF generator is handy for FM i-f alignment. Appropriate crystals are good for "band finding" when calibrating a home-brew receiver, prior to accurate calibration with the marker generator. Specific frequency crystals can be obtained for special purposes such as the i-f of transceivers at 5.0, 6.0, and 9.0 MHz; and will make possible accurate alignment with assurance and repeatability not possible with the average tunable signal generator.

Referring to the front-panel photo, use of the controls is fairly apparent. The generator is turned on by advancing the **output** control from fully counterclockwise. The **generator mode** switch selects the desired signal. In the HF and LF modes, a crystal is plugged into the appropriate socket. The crystal oscillators require no adjustment or tuning and will work with any active crystal within their range. In the **af** mode, the **audio freq** control is adjusted to the desired frequency. Accurate calibration of this control was not attempted, since approximate frequencies in the audio range (e.g., 400 Hz and 1000 Hz) are usually adequate.

However, a few adjustments are required when first setting up the Gentrac. The Marker Generator requires zero beating the 100 kHz and 1 MHz outputs against WWV, by adjusting C2 and C3. After this, L1 and L2 should be tuned for maximum output. You can use the tracer with the rf probe plugged into the generator output and observe the peak on the VU meter, but the **af** mode switch must be in the **mod** position. Adjust the output and the tracer gain for a suitable reading. The approximate af frequency markings for the **audio freq** control may be spotted by heterodyning the output against a borrowed audio generator or a frequency test record on the hi-fi system.

A typical application for the Gentrac in aligning a communications receiver is as follows:

1. **I-f Alignment.** Set the **generator mode** switch to LF and a 455 kHz crystal in the LF socket. Set the **af mode** switch to **mod** and the **audio freq** control at 0.4 (400 Hz). Connect the generator output cable to the converter stage grid, and the tracer input lead to the receiver audio output. Set the tracer input switches to **low impedance** and **low gain**, and the tracer output to VU. Set the generator output and tracer input controls for suitable VU reading, and peak the i-f transformers.

2. **Dial Calibration.** Set the **generator mode** switch to 1 MHz. Connect the generator output to the receiver antenna input, and adjust to a suitable level. Adjust the receiver trimmers for markers at 1 MHz dial markings.

3. **Rf Alignment.** Set the **generator mode** switch to **noise.** Connect the generator output to the receiver antenna input. Connect the tracer input to the receiver audio output. Turn generator level control fully clockwise. Set the tracer gain control for suitable VU meter reading. Peak the rf and converter stage trimmers for maximum noise level on the VU meter.

Many other uses for the Gentrac will suggest themselves. The tracer is conventionally used for locating signal loss or distortion, using the appropriate probe, when receiver troubleshooting.

With a telephone pickup coil plugged into tracer input No. 1, you can have a handy telephone amplifier. And for a little code practice, patch the generator output to the tracer input, plug a Morse key into J2, and set the **generator mode** switch to af. When used as a practice oscillator, square-wave output may prove less fatiguing than sine wave, and you can adjust the af frequency to your own liking. Incidentally, J2, which keys the af oscillator, is mounted on the back of the minibox and may be excluded if not required.

150 MHz Frequency Synthesizer

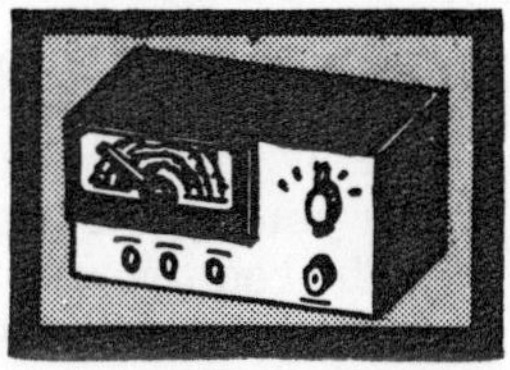

Peter A. Stark

First, we will give a complete description of what the synthesizer is and what it can do, along with a block diagram and simple description of how it works. If you decide to go ahead with building it, there's a complete parts list at the end of this installment. Next, we give the diagrams and the detailed theory of operation. Then we will give the printed-circuit board layout, parts layout drawings, and construction and operating information.

And now a little information about the unit:

The complete price should be under $100, depending on where you get your parts. We will have some information on parts sources in the parts list.

The unit covers from 145 to 148 MHz in steps of 5 kHz. With a few modifications the range could be extended down to 144 MHz, but that complicates the switching, so we decided against it.

It is inherently quite stable. It has only one crystal, at 10 MHz. Once you beat this crystal against WWV, all other frequencies automatically fall in line.

The synthesizer can be used for both transmitting and receiving, with a choice of 2-frequency operation. That's a necessary feature for working through repeaters, since you can transmit on one frequency and receive on another.

It is designed as a crystal substitute; you merely plug the synthesizer output cables into the crystal sockets of your rig. It is not a complete transceiver by itself.

The synthesizer is designed for receivers having either a 10.7 or 11.7 MHz i-f with low-side first-oscillator injection.

Since the synthesizer is only a crystal substitute, a regular transceiver is used for all other functions (such as modulation). But in case you have an AM rig, it is possible to frequency modulate the synthesizer itself. That will give you an FM signal. In our prototype we had a little extra room in the case, so we decided to install a touch-tone pad, which modulates the synthesizer.

We went out of our way to make the project easy to build. Except for the power supply, switching, and output buffers and multipliers, everything fits onto a 7 x 9 in. printed-circuit board. The PC board layout and etched boards are available to make your job even easier.

The requirement that the synthesizer work with a variety of transceivers makes the design a bit awkward, since different transceivers use different crystal frequencies. You can ignore the circuit capacity that your unit is designed for, since that has no effect on the output frequency with the synthesizer, but the crystal frequency itself (and the equation used to calculate it) is an important factor. Table I lists the requirements for several transceivers. As you can see, the actual crystal frequencies range from 6 MHz to 45 MHz. Most of the rigs use a 10.7 MHz receive offset, but the Standard uses 11.7 MHz; this requires a small change to the synthesizer. (The new Swan FM 1210-A uses 16.9 MHz offset. You'd have to make a number of changes to adapt the synthesizer to that rig.)

As you can see in Table I, most of the transmit crystals are 8, 12, or 18 MHz (the older Regency HR-2 uses 6 MHz rocks). We decided to build our unit for output at 6 MHz, which can easily be doubled to 12 or tripled to 18 MHz. If you need 8 MHz, you will have to change the design a little to "output" at 8 MHz directly. It is also possible to modify the synthesizer to "output" directly at 12 MHz, but it's easier to take the 6 MHz output and double it.

As for receiving, most rigs use 45 MHz rocks. You can start with 6 MHz and multiply by 8, start with 8 MHz and multiply by 6, or start with 12 MHz and multiply by 4. The 15 MHz signal needed by the Standard can be obtained from 8 MHz by doubling. But keep in mind that if you plan to use the synthesizer for both receiving and transmitting you have to pick a common frequency output. You can't pick 6 MHz output if you need 15 MHz for the receiver, since you can't double or triple 6 to get 15. Since 6 MHz seems to be the most useful with the widest variety of units (including our Varitronics IC-2F and an old Pre-Prog transmitter) we decided to design for that. The rest of this discussion describes the 6 MHz version, but we will tell you the applicable information for the 8 or 12 MHz versions. The additional information will always be in parentheses, following the 6 MHz information.

Crystal Frequencies for Common Transceivers					
Unit	Transmit Formula	Freq.	Receive Formula	Freq.	Synthesizer Output
Varitronics IC-2F	$\frac{f}{8}$	18 MHz	$\frac{f-10.7}{3}$	45 MHz	6 or 12 MHz
Drake	$\frac{f}{12}$	12	$\frac{f-10.7}{3}$	45	6 or 12
Standard SCR-826M SRC-146	$\frac{f}{18}$	8	$\frac{f-11.7}{9}$	15	8
Swan FM-2X	$\frac{f}{12}$	12	$\frac{f-10.7}{3}$	45	6 or 12

PRINCIPLE OF OPERATION

And now, on to details. Figure 1 shows the block diagram of the overall unit. The individual diagrams for each block will be shown next month; the block diagram gives the figure numbers for each detailed diagram.

The heart of the unit is the voltage-controlled oscillator, which we will simply call the **VCO** from now on. The VCO actually generates the 6 MHz (or 8 or 12 MHz) output frequency. This signal is fed into a switched buffer amplifier and then into the frequency multipliers needed to generate the correct crystal frequencies for your rig. Two sets of multipliers are usually needed, one set for the transmitting frequencies and one for the receiving frequencies.

All the rest of the synthesizer is used as an automatic frequency control to keep the VCO on frequency. The correct terminology is actually to call it a phase-locked loop (PLL). This is done by a feedback loop that compares the VCO output frequency with a crystal-generated reference frequency in the phase (and frequency) comparator. It is used here as a frequency comparator, but since it works by actually comparing phases, it is called a **phase comparator**.

Since the crystal frequency and the VCO output frequency are not the same we can't really compare them directly. Instead, we take the 10 MHz crystal frequency and divide it by a factor of exactly 48,000 in the reference frequency divider to get 208-1/3 Hz (in the 8 MHz version we divide by 36,000 to get 277-7/9 Hz, and in the 12 MHz version we divide it by 24,000 to get 416-2/3 Hz).

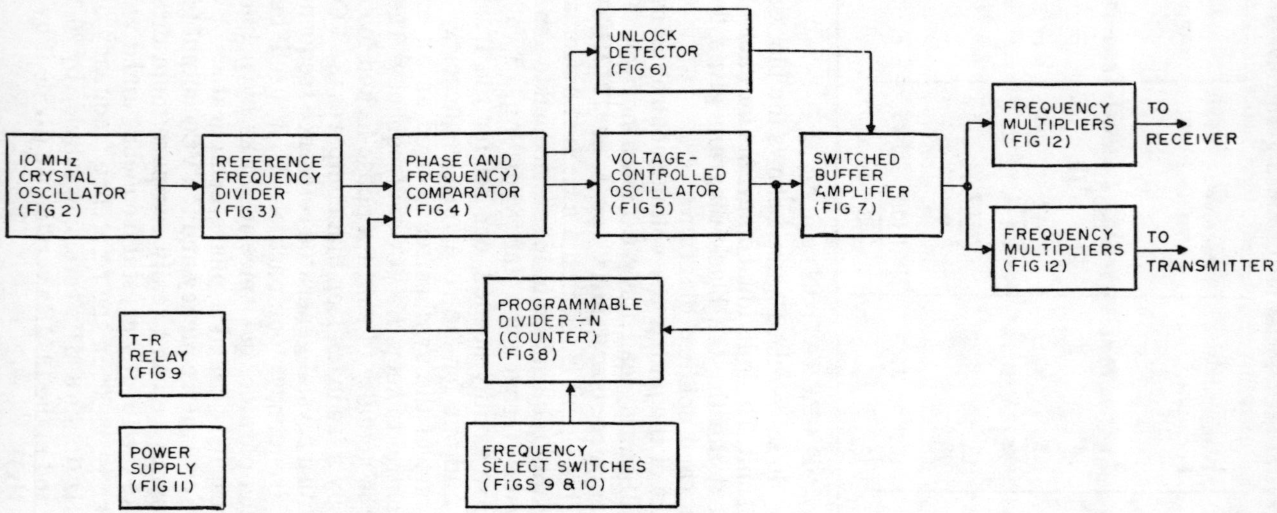

Fig. 1. Frequency synthesizer.

At the same time, we take the VCO output frequency and divide it by another number, **N**. But this division is done in a programmable divider, and the exact number by which we can divide is changed by the **frequency select** switches. Suppose we want an output on 145.000 MHz. Then the switches are wired in such a way that the programmable divider divides by exactly 29,000. Then the divider takes the VCO output frequency, divides it by 29,000, and sends the result to the phase comparator.

The phase comparator "looks" at the two frequencies it gets. If they are exactly the same, nothing happens. But if the frequency coming out of the programmable divider is too low, the phase comparator decides that the VCO frequency is too low also, and sends a control signal to the VCO to increase its output frequency. In the same way, if the frequency fed back from the programmable divider is too high, the phase comparator forces the VCO frequency down. In this way the phase comparator controls the VCO so as to make the output from the programmable divider the same frequency as the output from the reference frequency divider. This is the automatic frequency control action.

Now, if the output of the VCO, divided by the number **N**, is equal to the reference frequency, another way to look at it is that the output of the VCO is the reference frequency **times N**.

Using the 6 MHz synthesizer as an example, if we set the switches for 145.000 MHz output (so that N is 29,000), then the output from the synthesizer is the reference frequency (208-1/3) times N (29,000). If you multiply this out, you get exactly 6.041667 MHz. This is the exact frequency you would need to multiply up to 145,000 MHz. (If you want to do a little work, you can multiply 29,000 by the reference frequencies used in the 8 or 12 MHz synthesizer, to check that you get the right output frequency.)

Since the phase-locked loop forces the VCO output frequency to equal the reference frequency times **N**, the way we get various output frequencies to cover the range from 145 to 148 MHz is to change **N**. This is done with eight front-panel frequency selection switches. Just as an example, at several typical frequencies we have the following:

At 145.000 MHz, **N** is 29,000.
At 145.005 MHz, **N** is 29,001.
At 145.010 MHz, **N** is 29,002.
At 146.940 MHz, **N** is 29,388.
At 147.995 MHz, **N** is 29,599.

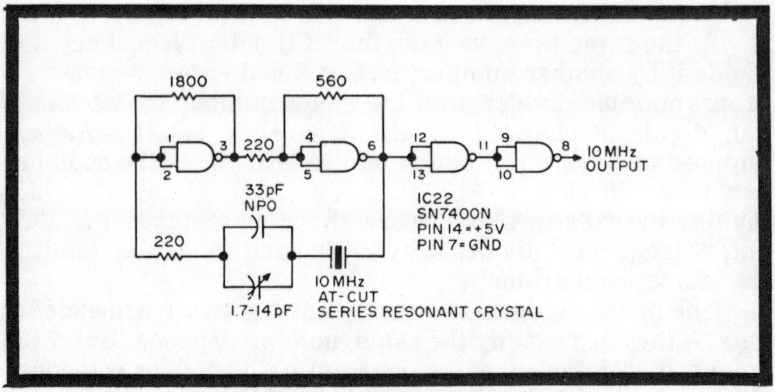

Fig. 2. The 10 MHz crystal oscillator.

As you can see, changing **N** by 1 changes the output frequency by exactly 5 kHz, thus giving us the 5 kHz channel spacing. If we make the programmable divider adjustable so **N** goes from 29,000 to 29,599, that will cover a 3 MHz band.

Since each count in the programmable divider changes the output by 5 kHz, if we subtract 2140 from **N** during receiving, we will automatically drop the output by 10.7 MHz. (To use an 11.7 MHz receive i-f offset, we subtract 2340.) This subtraction is done automatically by a TR relay. In fact, the switches can be wired to indicate the actual channel frequency. It is not necessary to figure out the value of **N** corresponding to the operating frequency. If you want to transmit on 146.340, just set the four **transmit** switches to the digits 6340. To receive on 147.015, just set the four **receive** switches to 7015. It's that simple.

When you use the synthesizer for transceiving, the VCO has to switch back and forth between two frequencies every time you push or release the mike button. As with any feedback system, this takes a little time. In the synthesizer, it takes the VCO about a quarter to a half second to reach and settle on the new frequency. To make sure that you don't transmit on the wrong frequency during this interval, an **unlock detector** monitors the phase detector output, and turns off the switched buffer amplifier while the VCO is moving to the new frequency. This makes it a little rough to break into conversations, but we've researched the problem and, short of having separate synthesizers for receiving and transmitting, there seems to be no way of getting around the problem. We could reduce this time below a quarter of a second, but only by greatly complicating the unit and upping the cost.

10 MHz CRYSTAL OSCILLATOR

The basic reference for the synthesizer is an AT-cut series resonant 10 MHz crystal in a simple circuit using the four gates from an SN7400N IC (see Fig. 2). This is a fairly simple circuit that is stable enough for the purpose. If you want the utmost in stability, though, you may want to put the entire circuit in an oven. The crystal oscillator circuit is mounted at the very edge of the PC board, and so is easily removed later, if necessary.

REFERENCE FREQUENCY DIVIDER

The frequency divider (Fig. 3) takes the 10 MHz signal from the crystal oscillator and divides it down into the reference frequency needed by the phase comparator. In our version, which provides 6 MHz output, the reference frequency divider provides a reference frequency of 208-1/3 Hz. For a synthesizer output of 8 MHz, the reference would be 277-7/9 Hz, and for a 12 MHz output the reference would be 416-2/3 Hz. The division from 10 MHz down to these frequencies is done by five ICs. The note at the bottom of Fig. 3 tells the differences in hookup for the three different circuits.

PHASE COMPARATOR

The reference frequency and the output from the programmable divider both go into the phase comparator,

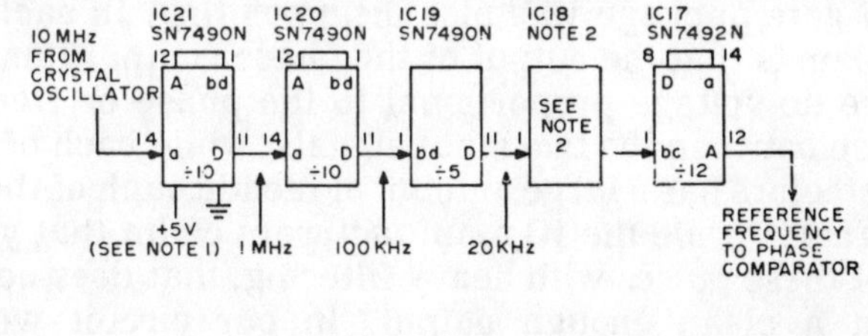

Synthesizer Output Freq.	Type	IC18 Data Divides By	Output Pin	Output Freq (Hz) from IC18	from IC17
6 MHz	SN7493N	8	11 (D)	2500	208-1/3
8 MHz	SN7492N	6	8(D)	3333-1/3	277-7/9
12 MHz	SN7493N	4	8(C)	5000	416-2/3

Fig. 3. Reference frequency divider. All ICs shown have +5V to pin 5, and ground to pins 2, 3, 6, 7, and 10; IC18 type and wiring depends on desired synthesizer output frequency, as shown.

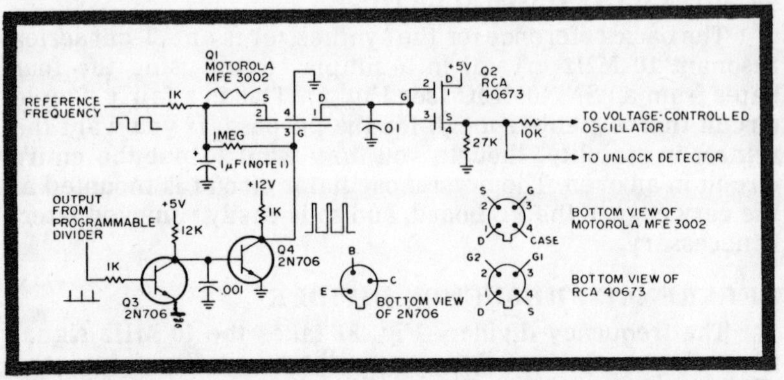

Fig. 4. Phase (and frequency) comparator. For 8 MHz, change to 0.75 uF; for 12 MHz, change to 5 uF.

(Fig. 4) where they are compared. If there is any difference in the two, a control voltage to the VCO changes the VCO frequency to bring it back to the right frequency.

We tried many circuits before coming up with the one in Fig. 4. There are several integrated circuits specially designed for phase-locked loop applications, and we tried them all. Signetics makes a series of phase-locked loop ICs that contain the phase detector, several amplifiers, and the VCO, all in one, but we found the performance unsatisfactory. Motorola makes a phase comparator IC, which also did not work out well. An exclusive-OR circuit made out of an SN7400N gate package was also disappointing. In each case, the problem is that the output of the phase comparator has to be a pure dc voltage proportional to the phase or frequency difference between the two input signals, while each of the IC phase detectors has a large amount of feedthrough of the input pulse signals. While the IC manufacturers claim that you can get rid of these pulses with heavy filtering, that does not quite result in a clean enough output. In our circuit we have managed to eliminate all traces of the input pulses on the output line.

The reference frequency input into the phase comparator is a square wave, which is applied to an integrator circuit consisting of the 1K resistor and 1 uF capacitor. These two components shape the square wave into a somewhat distorted triangular wave (with curved edges), which is then applied to the source of an MFE 3002 FET. This is an enhancement-type MOSFET that acts like a very fast switch.

At the same time, the output from the programmable divider is applied to the base of Q3. The signal coming in is a series of very narrow pulses (on the order of 0.1 usec) whose repetition rate is equal to the reference frequency (when the synthesizer is operating normally). The 0.001 uF capacitor in the collector circuit of Q3 widens these pulses to about a microsecond, so that the output of Q4 consists of wider pulses. Between pulses, the output of Q4 is about 0V, and each pulse goes up to about +12V.

These positive-going pulses are connected to the gate of Q1, which forms what is called a **sample-and-hold** circuit. When its gate is held around 0V (between the pulses from Q4), Q1 acts like an open switch. But during the 1 usec pulses, when the gate of Q1 is at +12V, Q1 acts like a closed switch. Hence, for a very short instant during each pulse, Q1 connects the 1 uF and the 0.1 uF capacitors together, charging the 0.1 uF capacitor to the exact voltage that happened to be on the 1 uF capacitor at the instant the pulse occurred.

If the pulse frequency output of the programmable divider is exactly the same as the reference frequency, then the pulses always occur at the same time in relation to the triangular wave. In that case, the voltage across the 0.1 uF capacitor will always be the same, since the FET switch will always sample the same part of the triangular wave. But if the two frequencies are not the same —so that the pulses occur either more often or less often—then the FET switch will sample different parts of the triangular wave, and the voltage across the 0.1 uF capacitor will change.

The output of the 0.1 uF capacitor is connected to a source follower, using an RCA dual-gate-protected MOSFET, the 40673. The source follower is very similar to a cathode follower, and provides a very high input resistance to the 0.1 uF capacitor. The purpose here is to monitor the voltage across the 0.1 uF capacitor without loading it down, so that the voltage across the capacitor stays constant even between successive input samplings—this is what is meant by "sample-and-hold." The output from the source follower is then sent to the **unlock detector** and to the VCO.

As pointed out earlier, the output from the phase detector is a constant dc voltage when the synthesizer is operating normally. This dc voltage is in the range of about 1.9 to 4.4V, and is used to control the frequency of the VCO. Under normal operation, the loop is locked on the reference signal. But if the phase detector output is changing, that means that the two inputs into the phase detector were not the same frequency, and thus the VCO frequency is changing. Under these conditions, the loop is not locked, and the unlock detector sensed this by seeing that the output of the detector is a changing voltage.

VOLTAGE-CONTROLLED OSCILLATOR

The VCO (Fig. 5) is another tricky part of the system, and we must confess that we are not quite happy with this one yet. The heart of the VCO is a Motorola MC1648P ECL oscillator IC, which uses L1 and C1 to set its frequency. The output of the IC is amplified and conditioned by a 2N5771 transistor, and then further amplified and buffered by some gates before being sent to the rest of the system.

The values specified for L1, C1, and C2 in Fig. 5 are for 6 MHz operation only; we haven't built the 8 or 12 MHz versions yet, so the choice of coil and capacitor for other operating frequencies will be up to you. We will discuss in a moment the way to decide these values.

The oscillator is tuned over its range by changing the shunt capacitance across the tuned circuit. Instead of using a varicap, we found two paralleled 1N4001 diodes just as good. The VCO control voltage (which is positive) is applied to the cathodes of these two diodes. The higher this positive voltage, the lower the diode capacitance, and the higher the operating frequency.

Because the oscillator has to operate at a lower frequency during receiving, we switch in another capacitor, C2. This is done by applying +5V to the base of Q5, turning it on. When Q5 is off, C2 is essentially out of the circuit, except for about 5 pF circuit capacitance.

Since the phase comparator output ranges from about 1.9 to 4.4V, we designed the VCO to cover the entire range with a control voltage from 2.2 to 4.1V. This leaves a 0.3V overlap at both ends, to make sure that enough control voltage is available from the phase detector to control the VCO over the whole range. On our 6 MHz synthesizer, the output frequency during transmitting is in the range of 6041 to 6167 kHz, and during receiving, it is in the range of 5595 to 5721 kHz. Hence, our VCO is designed to satisfy the data in Table 2.

Table 2. Data for 6 MHz Synthesizer

	Output Frequency from VCO (kHz)	
Input Control Voltage	Transmit (0V on base ofQ5)	Receive (+5V on base of Q5)
+2.2V	6040 or less	5595 or less
+4.1V	6170 or more	5725 or more

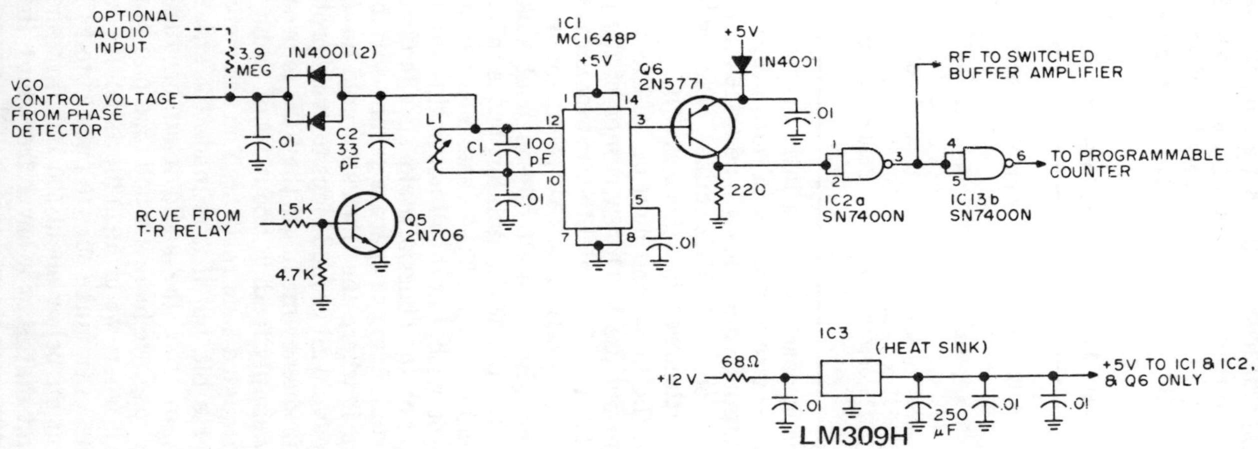

Fig. 5. Voltage-controlled oscillator. Coil L1 is 26 turns of no. 32 enamel wire close wound on 7 / 32 in. dia. by 19 / 32 in. long ceramic core, with ferrite slug (6 MHz output). Coil is adjustable from 3 to 6 uH, and is normally set to about 4.8 uH (6 MHz output).

If we build the synthesizer with 8 MHz output, then the transmitting output frequency range is from 8055 to 8222 kHz, and the receiving range (assuming 10.7 MHz i-f offset) is 7461 to 7628 kHz. Then we will have to change L1, C1, and C2 to get the data in Table 3.

Table 3. Data for 8 MHz Synthesizer

	Output Frequency from VCO (kHz)	
Input Control Voltage	Transmit (0V on base of Q5)	Receive (+5V on base of Q5)
+2.2V	8055 or less	7460 or less
+4.1V	8225 or more	7630 or more

Notice that in each case we add a little overlap, just to make sure that the final circuit is going to work properly. For 12 MHz output, the output frequencies will be twice those in the 6 MHz case. Also, for 11.7 MHz offset, the receiving frequencies will be a little lower than those shown.

In order to keep the noise and hum out of the VCO, we found it necessary to put the VCO in a corner of the board, separated by a ground strip from all other components on the board, and had to install a separate +5V regulator. We used an LM309H regulator with a small clip-on heat sink. The LM309H is similar to the LM309K regulator, but is in a small, TO-5 style can instead of the big TO-3 case. You can use the LM309K instead, if you wish, but it will not fit directly on the board. If you mount it on the chassis, make sure to insulate it from the chassis ground. Bring a separate ground wire back to the VCO and ground it where the LM309H would have been grounded. You must avoid all ground loops to avoid problems.

It is possible to frequency modulate the VCO to produce a very nice signal, by simply feeding about 1V of audio to the VCO control input through a 3.9M resistor. You can reduce the value of this resistor a bit, but if it is made too small, you probably will get hum from the audio input because of a ground loop. In our case, we have a touch-tone pad mounted on the case, and this is where we bring in the audio. If you want to voice modulate the VCO, make sure to put a cutoff filter in the audio path to cut off below about 500 Hz. Any audio introduced at this point that is at a lower frequency than the reference signal at the phase comparator will appear as an error in the phase-locked loop, and the loop will try to correct

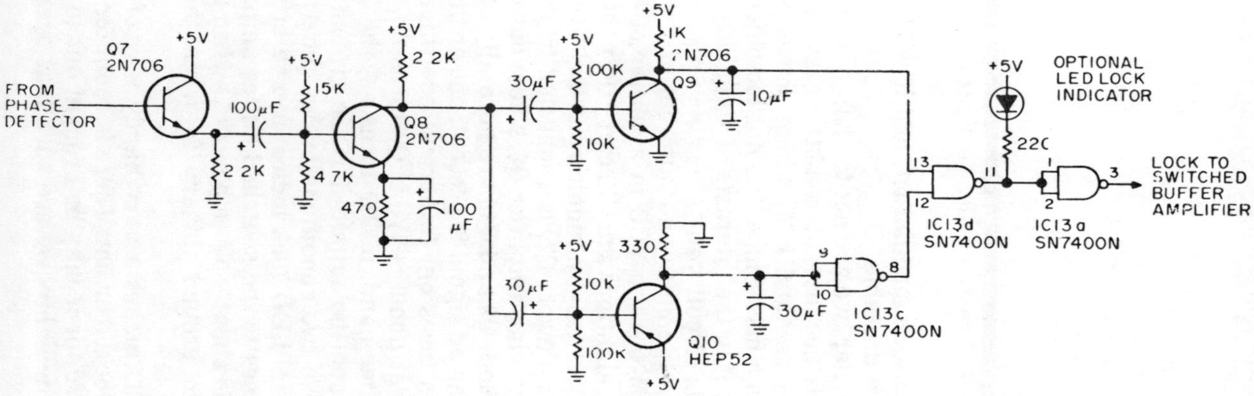

Fig. 6. Unlock detector.

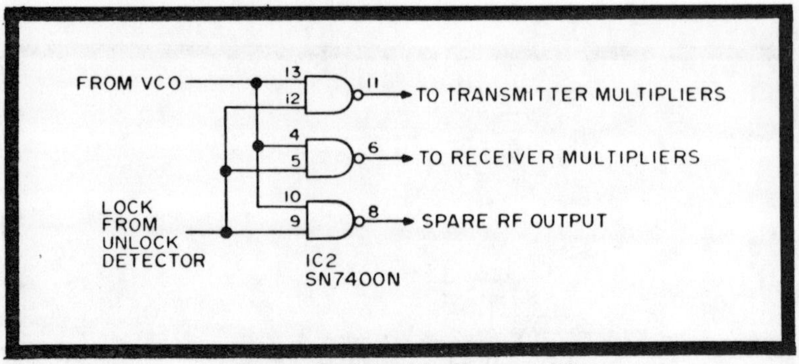

Fig. 7. Switched buffer amplifier.

for it. You won't get much modulation, and it will just make the loop a little more unstable.

As mentioned earlier, we are not too happy with the VCO at this point, primarily because it has a little noise. The output has a bit of random residual FM, which shows up on the transmitted signal as a hiss. Since it is not too objectionable, we decided to accept it in the interests of simplicity, but may get ambitious on this circuit yet.

UNLOCK DETECTOR

A relatively straightforward circuit (Fig. 6) checks whether the loop is locked by monitoring the VCO control voltage. An emitter follower, Q7, is connected to the control line, with its output driving amplifier Q8, whose output in turn goes to Q9 and Q10. Both Q9 and Q10 are normally biased just below turning on. If an ac signal is present, one or the other of these two transistors turns on. In normal operation, pins 13 and 12 of IC13d are at about +5V. If Q9 turns on, pin 13 is grounded; on the other hand, if Q10 turns on, the input into IC13c goes to +5V, and the output of IC13c ground pin 12 of IC13d. Either of these two conditions brings pin 11 to +5V, turning off the optional LED **lock indicator**, and bringing the lock signal, which goes to the switched buffer amp, to about 0V. This lock signal is about +5V (a binary 1) when the loop is in lock, and goes to ground (binary 0) when the loop is unlocked.

The optional LED makes a very handy way of checking whether the loop is locked or not. There is room for the diode right on the PC board, or it may be brought out to the front panel. When the diode lights, we are sure that everything is operating normally.

SWITCHED BUFFER AMPLIFIER

The switched buffer amp (Fig. 7) uses three 2-input gates in IC2 to buffer the output. When the lock signal from the unlock detector is a logical 1 (+5V), the gates are **on** and the rf signal is sent to the multipliers. If the lock signal goes to zero, the gates are turned off, and the output is disconnected.

PROGRAMMABLE DIVIDER

Although this circuit (Fig. 8) looks the most complicated, it is actually very easy to understand. Integrated circuits IC4 through IC7 are programmable decade up-down counters, which, along with IC12, do the actual counting. As mentioned earlier, the value of N—the number by which we divide—is a constant between 29,000 and 29,599 during transmitting, and is 2,140 less during receiving. Temporarily forget about what happens during receiving, and look at the numbers **29,000** and **29,599**. Since the digits **29** at the beginning are the same all the time, IC12 is permanently wired to count to 2, while IC7 is permanently wired to count to 9. The last three digits of **N** are entered into IC6, IC5, and IC4, in that order, by the 11 pairs of diodes connected to pins 15, 1, 10, and 9 of those ICs. (Pin 9 of IC6 is always grounded, since that bit is always a 0.) The actual connections to the diodes are shown later.

To see how the divider works, let's take an example. Suppose we want to transmit on a frequency of 146.940, which corresponds to a value of 29,388 for **N**. The digits **29** are permanently wired into IC7 and IC12, the **3** is set into IC6, and the digit **8** is set into both IC4 and IC5. In this way, the counter is preset to the starting value of 29,388. As soon as this is done, each input pulse from the VCO circuitry subtracts 1 from the counter. After exactly 29,388 input pulses, the counter gets down to a count of 00000, and it repeats the whole process all over again.

To understand all of the fine detail, we have labeled some of the lines in Fig. 8 with letters **A** through **F**. After the counter starts counting down toward 0, a number of things happen. First, IC12 gets down to a count of 0; when this happens, line C goes "high" (meaning **positive**, or about +5V). A short time later, when IC7, IC6, and IC5 all reach the count of 0, line F goes low (near 0V); therefore, line B goes high. Then IC4 reaches a count of 2, line A goes high. Thus, when the counter has gone from its original preset count down to a count of 00002, the output of IC8c goes low. This applies about 0V to the D-input of IC14, a so-called Type-D flip-flop; but the flip-flop does not react until the next input pulse. At this pulse the counter tries to count down to 00001. But at the same time,

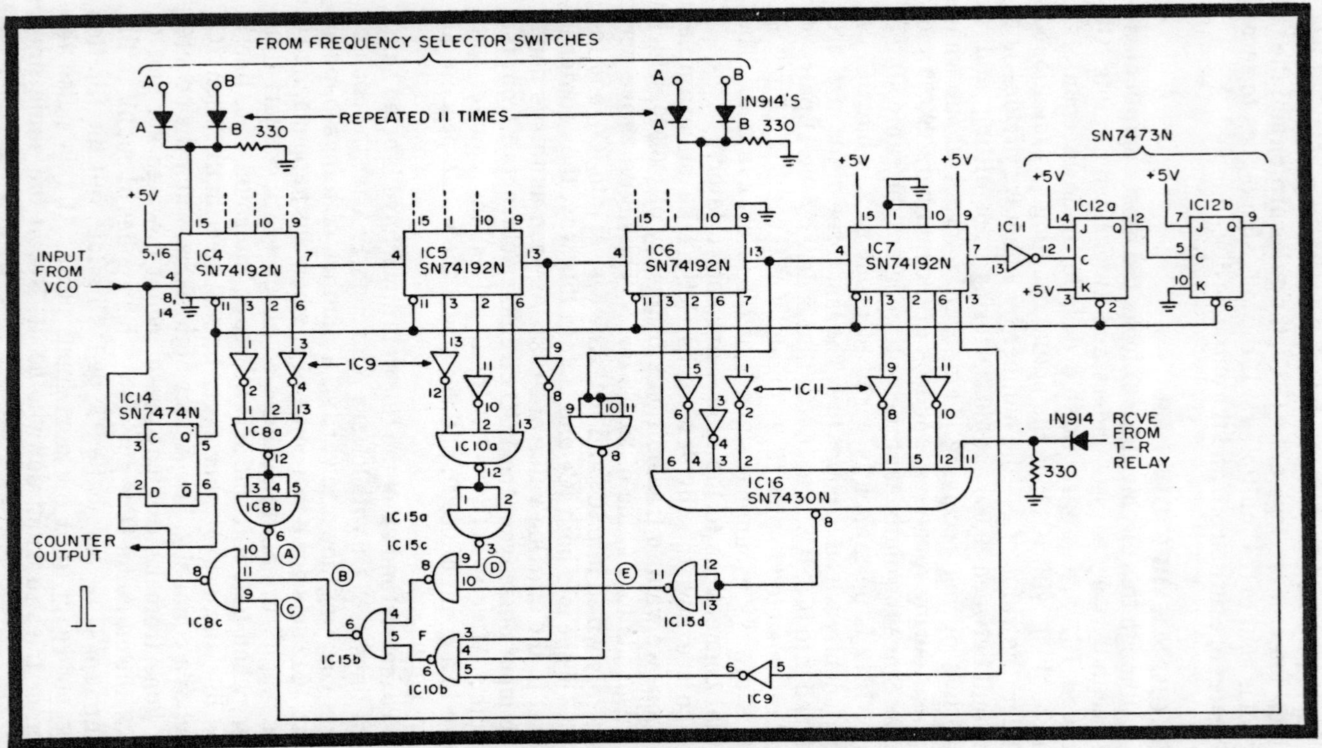

Fig. 8. Programmable divider (÷ N).

IC14, which was set until now, resets, and its Q-output grounds the load lines—pin 11 of IC4 through IC7, and pins 2 and 6 of IC12. This forces all of the counter stages to preset back to the starting value of 29,388. But the counter can't start counting yet—it has to wait until the next input pulse (when the counter would otherwise have been at the count of 00000), at which time IC14 sets again.

Hence, IC14 goes through a complete set-reset-set cycle exactly once every 29,388 input cycles (or whatever N corresponds to the frequency selected), which gives us the frequency division by N. The output from pin 6 of IC1; is a short positive pulse occurring exactly once every N input pulses.

When the synthesizer is switched into the receiving mode, the TR relay puts +5V on the "rcve" line, which, in turn, makes pin 11 of IC16 go high. This brings IC11, IC15, and IC16 into the picture. As before, the counter starts to count down from its preset value towards 0, but, this time, it never reaches anywhere near it. As soon as IC7 and IC6 reach a count of 21, line E goes high. When IC5 reaches a count of 4, line D goes high; so, line B goes high. Lines A and C work the same as before; so, the entire reset process starts when the counter gets to a count of 02142, and resetting finishes at the count of 02140 (instead of 00000 as in transmit). Since this reduces the quantity N by exactly 2140, it drops the frequency (after it is multiplied up to 2M range) by exactly 10.7 MHz.

If the first i-f of your rig is different from 10.7 MHz, it will be necessary to change this circuitry. For an 11.7 MHz i-f (such as in the Standard transceivers), for example, you need to start resetting the divider at the count of 02342 instead of 02142. In this case, the change is very simple—just eliminate the inverter (IC11 pins 5 and 6). Probably the best way to do this on the board is to cut off pins 5 and 6 on IC11s socket to disconnect the inverter, and then put a jumper between pins 5 and 6 on the board.

Notice also that we subtract the count of 2140 from the quantity N—we don't add it. That means we can only accommodate oscillator injection on the low side of the received signal. This fits most imported transceivers. But if you own the Yaesu transceiver, which uses high-side injection, the synthesizer changes required are really major, and it probably doesn't pay to use the available printed circuit board.

TR RELAY AND CHANNEL SELECTION SWITCHES

Undoubtedly we could have replaced the push-to-talk relay shown in Fig. 9 with a transistor, but the relay is more

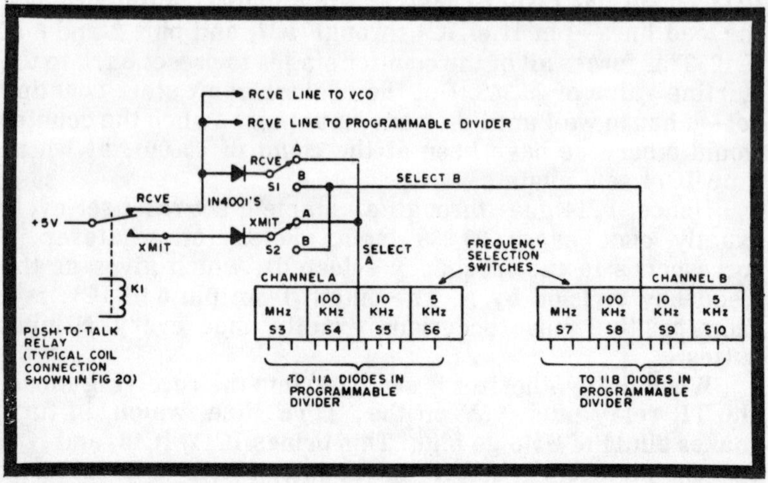

Fig. 9. TR relay and channel selection switches. (See Fig. 10 for detailed wiring of S3 to S10.)

readily adaptable to most transceivers. When the relay is deenergized, the synthesizer is in the **receive** mode, and +5V is supplied to the "rcve" channel switch, S1; to the VCO (to turn on the transistor, which switches extra capacitance into the VCO tuned circuit); and to the programmable counter (to cause subtraction of 2140 from the factor N). When the synthesizer goes to **transmit**, +5V is only supplied to the "xmit" channel switch, S2.

The "rcve" and "xmit" channel switches select either channel A or B for operation, and in turn send +5V to either the channel A frequency selector switches (S3 through S6) or the channel B frequency switches (S7 to S10). In this way we get independent frequency selection for transmitting and receiving, which you need for going through a repeater. For simplex operation (transmission and reception on the same frequency), you can also preset two different frequencies, and simply flip back and forth. With only a little more effort, you could add a scanning circuit that would automatically go back and forth between frequencies A and B; it would have to work slowly, though, since the synthesizer takes about a quarter to a half second to settle on a new frequency (it will take less time if the two receiving frequencies are close together).

Frequency Selection Switches

As shown back in Fig. 8, there are 11 preset inputs brought out from IC4, IC5, and IC6. Each input goes to ground through a 330-ohm resistor, and also goes to two diodes. Of each pair of diodes, one goes to the channel A frequency selection switches, and the other goes to the channel B switches. The diodes are there to isolate the two sets of switches from each other.

The actual wiring of these switches is shown in Fig. 10. The MHz switches (S3 and S7) are single-pole, 3-position rotary switches; the hundreds-of-kHz switches (S4 and S8) and the tens-of-kHz switches (S5 and S9) are all 3-pole 10-position

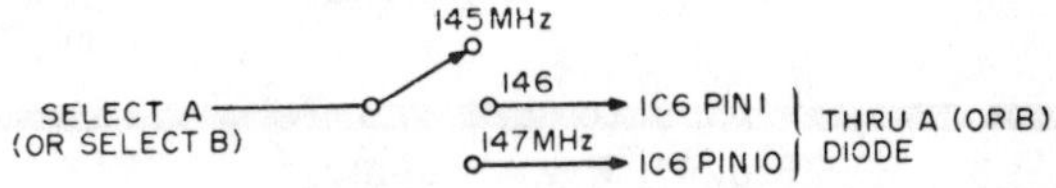

a. MHz selection switches S3 and S7.

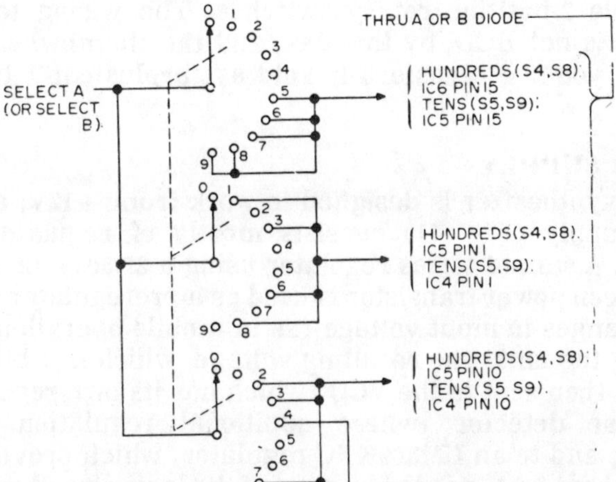

b. Hundreds (S4 and S8) and Tens (S5 and S9) switches.

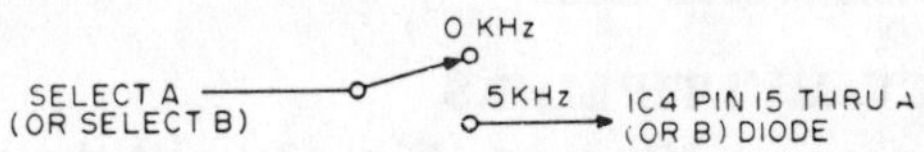

c. 5 kHz selection switches S6 and S10.

Fig. 10. Wiring of frequency selection switches S3 through S10.

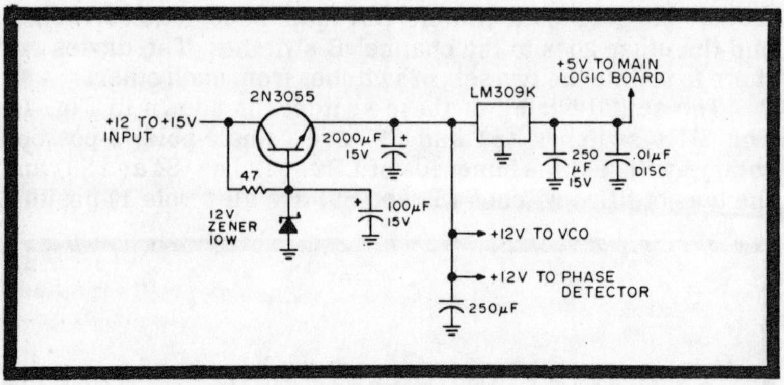

Fig. 11. Power supply.

rotary switches, and the 5 kHz switches (S6 and S10) are single-pole 2-position rotary switches. The wiring to these switches is not BCD, by the way, and the thumbwheel BCD switches you see occasionally sold as surplus can't be used here.

POWER SUPPLY

The synthesizer is designed to work from +12V, and the power supply (Fig. 11) consists mostly of regulators and filtering. A simple series regulator using a 2N3055 or similar NPN silicon power transistor is used as a preregulator to keep large changes in input voltage (as in mobile operation) from affecting the unit. The resulting voltage, which is a bit under +12V, is then sent to the VCO (which has its own regulator); the phase detector (where additional regulation is not needed); and to an LM309K IC regulator, which provides the +5V at about 0.6A, needed by most of the logic circuitry.

The LM309K in the main regulator should be mounted on a heat sink or on the chassis. The LM309H in the VCO circuitry mounts right on the board, and should have a small clip-on heat sink. The 2N3055 and the 12V zener should also be heat sinked or chassis mounted.

FREQUENCY MULTIPLIERS

The frequency multipliers (Fig. 12) take the output of the synthesizer and multiply it up in frequency to whatever value is needed by the transceiver. Since various transceivers need different amounts of multiplication, we decided not to design our own, but to simply use a commercially built amplifier, the PAX-1 power amplifier of International Crystal.

As shown in Fig. 12, the circuit is quite simple. As it comes in the kit, the amplifier is designed for class C and needs quite a bit of drive. We added a 33K resistor to provide some base bias (it mounts in a set of holes already provided on the board), and found that it works very well as a frequency multiplier. Though it is only rated as being good up to 30 MHz, we found that it works at 45 MHz. In our unit, we use four of these units: One is to triple from 6 MHz to 18 MHz, for our transmitter; three are to double from 5.6 MHz (during receiving) to 11.2 MHz, then to 22.4 MHz, and finally to about 45 MHz. The kit comes with three color-coded coils, which cover the range through 45 MHz. But disregard the instructions provided with the kit as to the number of turns in the link, and use the number of turns specified in Fig. 12 between multipliers. The last link, used to couple from the multiplier to your transceiver, will have to be chosen experimentally. If you have an rf voltmeter or high-frequency oscilloscope you are in luck, since then you can measure the voltage drive using a crystal, and can tailor the link to provide the same voltage.

CONSTRUCTION DETAILS

Most of the construction details are pretty well covered by the layout drawings, so the text itself will cover only some items. When looking at the drawings, make sure to read the fine print in the notes, which may explain some questions.

The main body of the synthesizer fits on the PC board, which is shown in Fig. 13. It is a little smaller than 7 x 9 in., and

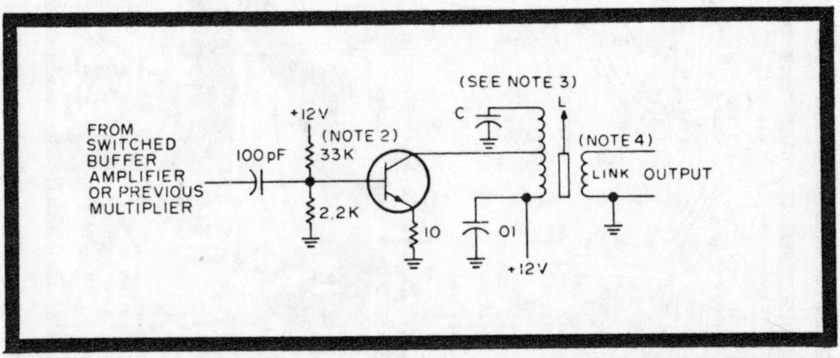

Fig. 12. Frequency multipliers. (Notes: 1. Use components and board of International Crystal PAX-1 rf power amplifier. 2. Add a 33K resistor into holes marked "R" on PAX-1 board. 3. Select L and C as shown in PAX-1 instructions for range covered. Adjust for max output with synthesizer set for 146.5 MHz. 4. Use 10-turn link with yellow coil and 5-turn link with red coil for coupling to next multiplier. Experiment with correct number of turns on final link to transceiver.

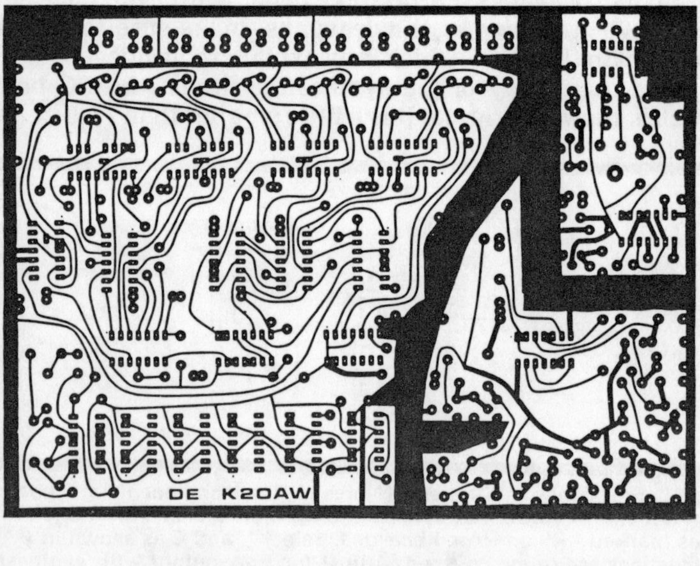

Fig. 13. Copper side of printed-circuit board (60 percent reduction).

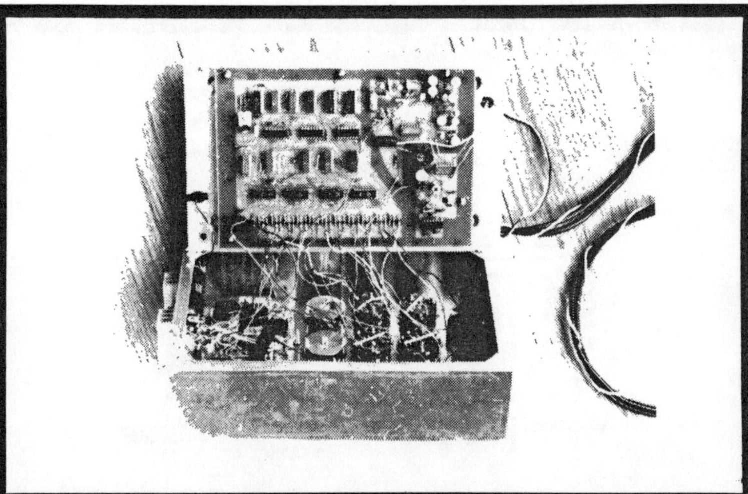

is etched on only one side, to make it easier to duplicate. This means that there are a lot of jumpers, so watch out, and don't miss any. You can use bare wire for the jumpers, since none of them crosses any other jumpers or components. Be careful, though, since some of the jumpers run under IC sockets.

In order to explain where all the PC board mounted parts go, we have broken up the main board into smaller sections, shown in Figs. 14 through 19. Each of these takes one or two of the main circuits shown above, and shows the actual layout of

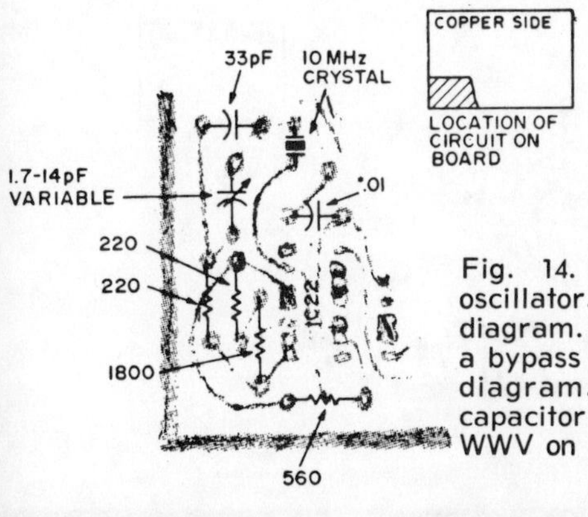

Fig. 14. The 10 MHz crystal oscillator. (**Notes:** 1. See Fig. 2 for diagram. 2. Capacitor marked is a bypass capacitor not shown on diagram. 3. Adjust variable capacitor to zero beat crystal with WWV on 10 MHz.)

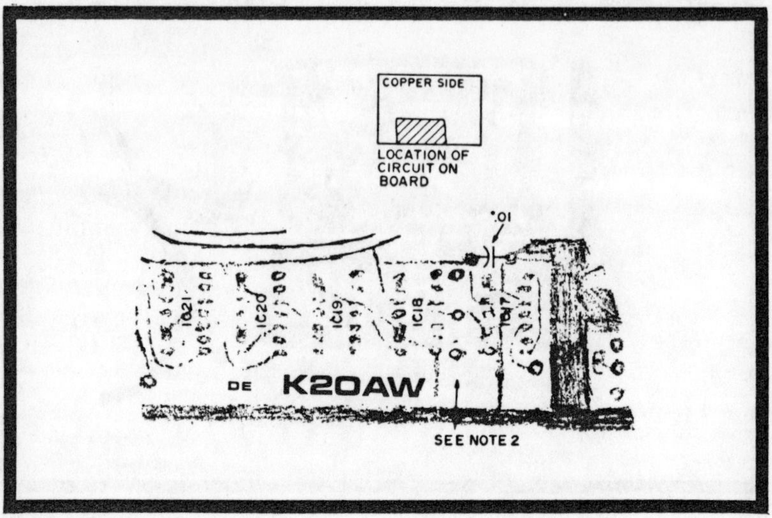

Fig. 15. Reference frequency divider. (**Notes:** 1. See Fig. 3 for diagram. 2. Jumper between IC18 output (pin 8 or pin 11) and IC17 input (pin 1) as shown in Fig. 3. 3. Capacitor marked is a bypass capacitor not shown on diagram.

the parts. In each drawing, a small sketch in the corner shows where to find that circuit on the overall board. In all of these pictures, we are looking at the copper side of the board. The components themselves actually mount on the other side of the board.

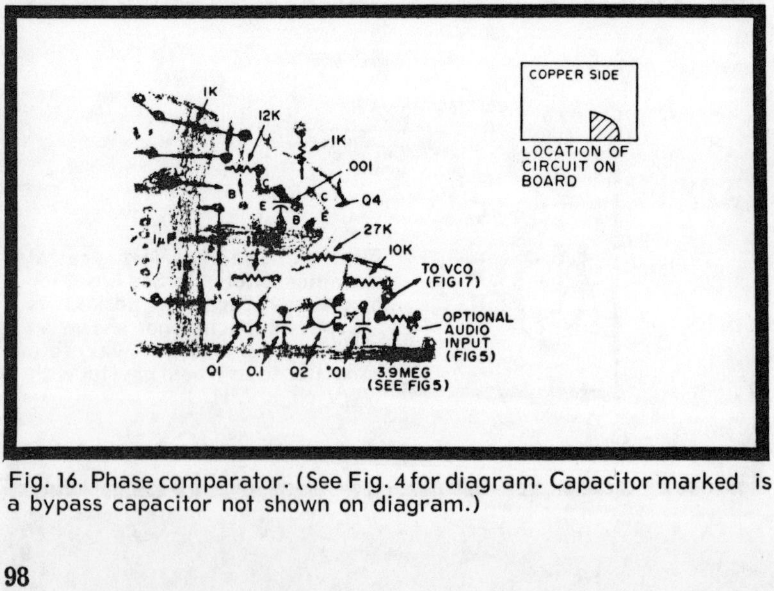

Fig. 16. Phase comparator. (See Fig. 4 for diagram. Capacitor marked is a bypass capacitor not shown on diagram.)

Orientation of the FETs is shown on the layout drawings. The letters **B**, **C**, and **E** indicate the way regular transistors plug into the board. The LM309H on the VCO section of the board plugs in (from the nonconducting side of the board, naturally) so that the three leads line up with the three holes. Note also that the other ICs have to be inserted in a certain way. There is usually a small notch or depression at one end of the IC body, on the end near pin 1. We have marked pin 1 of each IC on the board and board layout with a small dot.

For a number of reasons, IC sockets are a must for a project this size. Since IC sockets of the molded type are quite expensive (even more expensive than most of the IC's themselves) we suggest something called **Molex Soldercon IC terminals**, or just Molex pins. These are individual little clips, one for each IC pin, which come in long strips. You break off as many as you need (such as 7 for one side of a 14-pin IC), insert the whole strip into the board, solder, and then snap off the connecting strip. This leaves seven independent little clips standing up from the board. Put another seven on the other

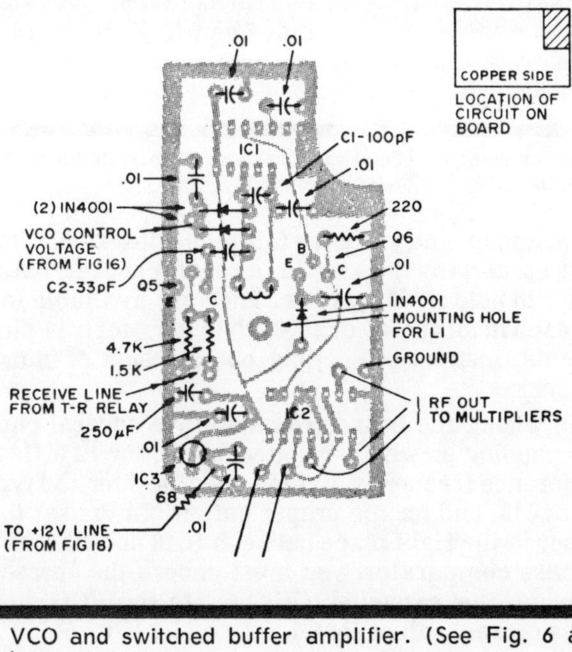

Fig. 17. VCO and switched buffer amplifier. (See Fig. 6 and 7 for diagram.)

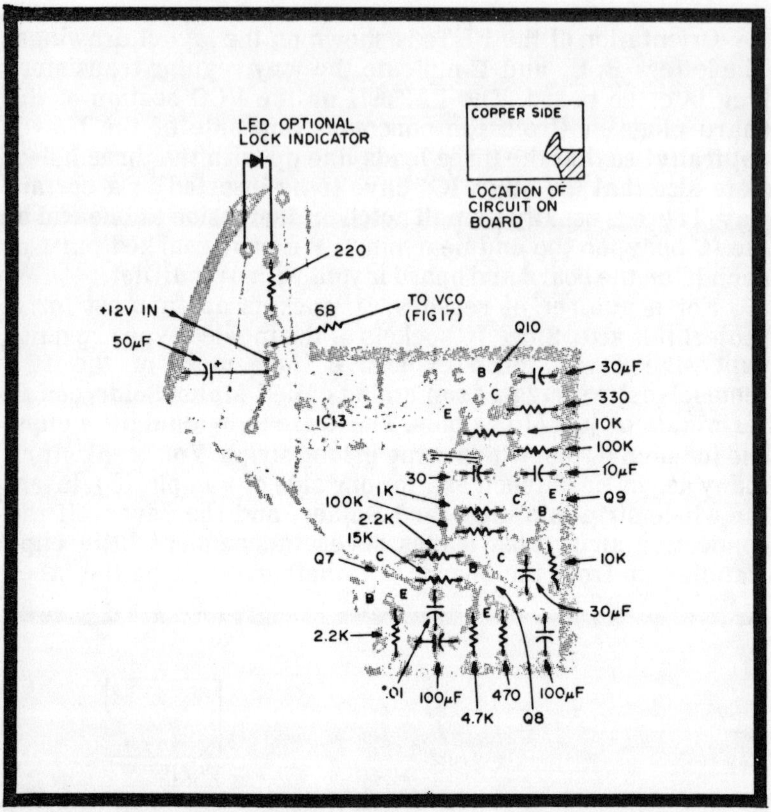

Fig. 18. Unlock detector. (See Fig. 6 for diagram. Capacitor marked is a bypass capacitor not shown on diagram.)

side of the lineup, and you have fourteen clips, ready for the IC itself. It's essentially the same as a regular socket, but without the plastic to hold it all together. They are available for about a penny each in lots of 500 or 1000; thus, you get a 14-pin socket for 14 cents, instead of shelling out 50 cents or more for a regular socket.

While wiring the board, you will have several choices to make, depending on whether you want 6, 8, or 12 MHz output. In the reference frequency divider, see Fig. 3 for the type of IC to use for IC18, and for the proper output pin to use; then, put the jumper in the right place between IC18 and IC17 (Fig. 15). In the phase comparator, you must choose the correct value for the integrator capacitor (Fig. 4). In the VCO, you must choose the correct values for C1, C2, and L1 (Fig. 5).

If you intend to use the synthesizer with an 11.7 MHz i-f in the receiver, make the change in the programmable divider as described earlier.

FET Precautions

Special precautions are needed in installing the two FET transistors. The RCA 40673 is gate protected, so it is relatively free from burnout due to careless handling; but we did manage to ruin one anyway, so it pays to be careful with both. The problem is that the gate resistance on MOSFETs is so high that static electricity can induce very high voltages between the gate and the rest of the FET. As a result, it is very easy to exceed the breakdown voltage between the gate and the channel, and literally punch a hole in the insulation layer between them.

The FETs should be the last components mounted on the board. When removing the MFE3002 from its package, you will note that a metal ring is slipped over all the leads, shorting all the leads together. Do not remove this ring at first Instead, take some very fine bare wire (one strand from a length of zip cord is fine) and carefully wrap it around all the

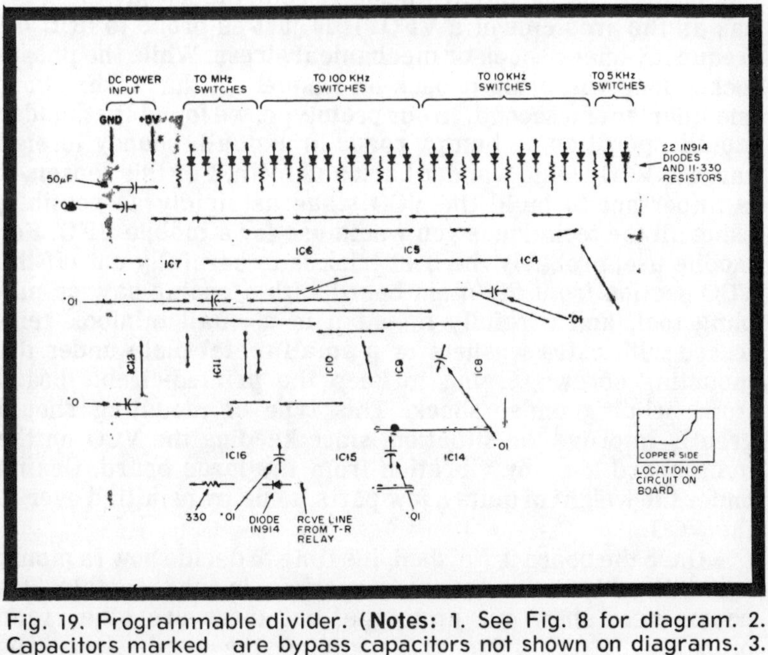

Fig. 19. Programmable divider. (**Notes:** 1. See Fig. 8 for diagram. 2. Capacitors marked are bypass capacitors not shown on diagrams. 3. Note the junpers running under IC4, 5, 6, 7, and 12.)

wires, up near the body of the transistor. Go around and in and out between the leads, to make sure that you have all the transistor leads securely connected together. Do the same with the 40673. Don't solder this wire to the leads, since it will be removed later. Now that the four leads from the FET are shorted together, you can remove the small ring supplied by the manufacturer, slip the leads into the PC board, and solder them. In soldering, use a small pencil iron (not a solder gun). Let the iron heat up, and then, just before using it to solder the FET, unplug the iron and connect a clip lead from the iron to the ground strip on the board. In this way you will prevent stray voltages from sneaking up on the FET and zapping it. Only after it is soldered do you remove the bare wire from between its leads. Once the FET is on the board, it is relatively safe, but try not to touch the FET case and circuitry anyway. These precautions are a little conservative, but we believe in playing it safe with these little critters.

Putting It All Together

When building the synthesizer, remember that the VCO has all the problems of a VFO. It is just as prone to drift off frequency under shock or mechanical stress. While the phase-locked loop will bring it back on frequency, that takes about one-quarter of a second. In our prototype, we found that, under mobile operation on bumpy roads in a small, bouncy foreign car, the VCO would wander back and forth. For this reason, it is important to build the VCO stage as rigidly as possible, using all the techniques you would use for a mobile VFO. For mobile use, probably the best trick is to carefully cut off the VCO section from the main board with a coping saw or nibbling tool, and carefully mount it in a small minibox reinforced with extra washers or a small metal plate under the mounting screws, trying to keep the printed-circuit board from bending under shock. This type of mounting should greatly improve the situation, since keeping the VCO on the main board lets any vibration from the large board, flexing under the weight of quite a few parts, to be transmitted over to the VCO.

Once the board is finished, it's time to decide how to mount it and the switches and other components in one assembly. The photos here show our prototype, but there are other (and probably better) ways of packaging the synthesizer. The important thing to keep in mind is that the overall assembly should be such that you can get to the board for testing or repairs, if needed. Wire the power supply regulator next, since it is a good idea to test the board before continuing further.

For testing purposes, connect the power supply regulator to the board. The +5V and ground connect to the corner of the board, as shown in Fig. 19. The +12V connects near IC13, as shown in Fig. 18, and goes to the VCO through the 68-ohm resistor (Figs. 17 and 18). A good source of power for the regulator is a 12V car or motorcycle battery.

Before applying power, check the resistance between the +5V and +12V lines to ground. If there is a short circuit, look for a wrong connection or tiny sliver of solder on the board. Don't forget to remove the Molex connecting strip before applying power, since it shorts adjacent terminals.

Checkout

From now on, a frequency counter is probably the best test instrument to use. We used the one described in **73 Magazine** in the May-September 1972 issues. If you can borrow one, it is well worth the effort, since, even if you don't need it in troubleshooting and everything works well, at least you will be able to check the accuracy of the synthesizer's output at different switch settings.

Connect the counter to the collector of Q6 in the VCO. Disconnect the VCO control line from the phase comparator, and connect it to a source of variable dc voltage, adjustable over the range of about 1.5-5.0V. Check that the output frequency of the VCO at the various input voltages shown in the VCO description is in the right range.

If necessary, adjust L1 so that the VCO gives the right output. The important thing is to check that the VCO frequency swings over the whole range needed with an input voltage change from about +2.2-4.1V, both on receive and transmit. If you don't get enough swing with this amount of voltage input, try another pair of 1N4001 diodes in the control line. It is possible to get a diode that does not have a large enough capacitance range, although only one diode out of all the ones we tried gave us trouble in that respect. If this does not help, then you will have to experiment with the values of C1, C2, and L1 until you get the right VCO frequency swing. This part of the synthesizer must be aligned first, or you may never get the rest of it to work. Only if the VCO tracks right can the rest of it lock onto frequency.

Now, disconnect the variable voltage source from the VCO control line, and measure the output frequency. It should be somewhere near the lower edge of the required frequency swing. Check that the same frequency is also on pin 4 of IC4 (though there may be a slight difference because of loading changes as you move the counter). Then check IC5 pin 4, IC6 pin 4, IC7 pin 4, and, finally, IC12 pin 1. Each of these should have a frequency one-tenth of the frequency on the preceding test point. If you get the wrong reading, try interchanging ICs with others of the same number.

Since at this point you don't have the frequency selection switches connected yet, the programmable divider is dividing by exactly 29,000. Connect your counter to IC14 pin 6 to check the output frequency from the divider. It should be the VCO frequency, divided by 29,000, which probably makes it fairly close to the reference frequency.

Next, check the output of the 10 MHz crystal oscillator and the reference frequency divider. Figure 3 tells the frequencies to be expected at various places in the reference divider.

Still without the VCO control line connected, use a scope to look at the inputs to the phase comparator (see Fig. 4). The square-wave reference frequency is easily observed, but you will need a very good scope (with triggered sweep and good high-frequency response) to see the output from the programmable divider. (That's why a frequency counter is so useful: You can tell that the signal is there and measure its frequency, even though it is too narrow to see.) Then check the phase detector output. Since the loop is not closed, the phase detector output should be varying all over the place. Seen on the scope, it should look like a staircase that continuously shifts directions and slopes.

Now comes the big step. If everything is working so far, connect the phase detector output to the VCO control input with a wire jumper on the underside of the board. Make it as short as possible, and keep it tight against the board. The synthesizer should now lock on frequency. Since the frequency selection switches are still missing, you are set to transmit on 145 MHz. The frequency counter should now measure the correct frequency (145 MHz, divided by 24, 18, or 12, depending on whether your output is at 6, 8, or 12 MHz, respectively). The counter reading should be constant, with possibly a flicker in the rightmost digit. Listen to the output of the synthesizer on a general coverage receiver to make sure it is reasonably clean.

If the synthesizer locks on the correct frequency but has a lot of hum or buzzing on the signal, chances are that something is wrong with the phase detector. Put a scope on the output—it should be a pure dc level, a straight line right across the scope screen. If it has a slight sawtooth to it, it's possible that one of the two FETs is shot.

Once you have gotten this far, wire up the frequency switches and package the whole system in a cabinet or box. The outside edge of the printed-circuit board is ground. Mount the board on standoffs on a chassis or grounded metal plate so that the ground area of the board is grounded through the screws. (You may have to readjust L1 in the VCO once this is done.) Keep all the leads short, and keep all leads as far away from L1 and Q5 in the VCO as possible.

Now comes the tricky part. The VCO is very sensitive to ground loops and hum pickup. As you remember, a voltage change from about 2.2-4.1V (a total swing of slightly under 2V) causes the VCO to cover the range from 145 to 147 MHz. This is a swing of about 1.5 MHz per volt of control voltage, which corresponds to about 1.5 kHz /mV. All this means that if you have 1 mV of ripple or noise anywhere around the VCO, that will show up as 1.5 kHz of deviation in your signal. To keep your output clean and pure, you have to avoid all pickup—that includes ground loops, inductive and capacitive coupling, etc. This is an experimental process at best. Adjust the synthesizer for an output on some frequency for which you have a receive crystal in your transceiver.

Place the transceiver close by, and you will be able to hear the synthesizer output on your rig. If you hear a buzz or hum,

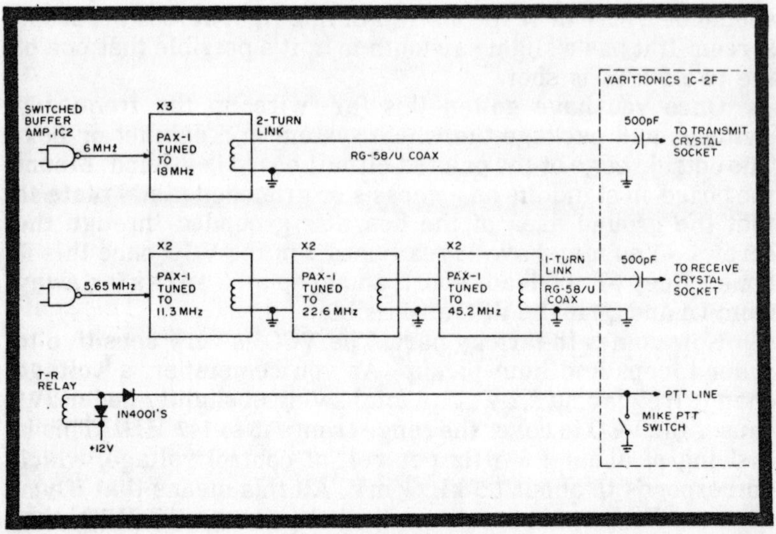

Fig. 20. Interfacing between the synthesizer and our Varitronics IC-2F.

you will have to try every trick you can think of to get rid of it. This may involve removing mounting screws holding the board to the chassis (in our unit, we found that we had to keep all mounting screws within 2 in. of the VCO insulated from ground); moving wires around; and checking ground connections between the power supply, switches, and relay.

This should be done with the synthesizer in the transmitting mode, either by putting +12V on the TR relay, or by bypassing the contacts. In our unit, placing our transceiver close by permits us to hear the 24th harmonic of the 6 MHz signal on the rig. Once you have checked the signal in the transmitting mode, repeat the check in the receiving mode. It is possible to hear the 26th harmonic of the 6 MHz signal, for example, by setting the frequency switches in the right position. If you set the switches to receive on 1;6.335 MHz, the 6 MHz synthesizer puts out a signal at 5.6514583 MHz (146.335, minus 10.7, all divided by 24). The 26th harmonic of this falls near 146.94 MHz and can be heard.

Tips on Using the Synthesizer

Once the unit is putting out a fairly clean signal on 6 MHz, you must interface it with your transceiver. This is done by using International Crystal PAX-1 rf power amplifier modules for doubling and tripling. If your synthesizer outputs on the frequency you need for your rig, then you should still use a

PAX-1 module as a straightthrough amplifier for buffering the output signal. This also puts a tuned circuit in the line, and changes the square-wave output of the switched buffer amplifier into a reasonably good sine wave. This points up the fact that it doesn't pay to build a 12 MHz synthesizer to provide a 12 MHz output, since either way you will need a PAX-1 module tuned to 12 MHz. You might as well follow our 6 MHz design, as is, and use the PAX-1 as a doubler.

Since the PAX-1 tuned circuit is fairly sharp, you will need separate multipliers for transmitting and receiving since the output frequencies in the two modes differ by about 10 percent. Figure 20 shows the interfacing between our transceiver, the Varitronics IC-2F, and the synthesizer. The output links were the same as those recommended by International Crystal for the coil-capacitor combination used in the final tuned circuit. We measured the amount of drive required by the transceiver at about 1-1.5V p-p, which was easily provided by the small links. The rf connections between the synthesizer and the transceiver were made by 52-ohm coax.

There are probably dozens of other hints we could give if only we could think of them at the moment. We have found the synthesizer an exciting project, since there is so little factual information in other books on the subject—and a lot of manufacturers' application notes have turned out to be misleading or **wrong**! We have spent literally hundreds of hours tracking down leads, doing calculations, and trying first one circuit and then another. We think you will enjoy building your synthesizer. At the time of writing, this is something you can't go out and buy—and that's the best kind of project to build.

PARTS LIST

4	SN7400N	TTL Quad two-input *nand* gates	1		1.7-14.1 pF air variable capacitor
3	SN7490N	TTL decade counters	2	33 pF	(Johnson 189-505-5)
1	SN7492N	TTL Divide by 12 counter	1	.001	NPO disk capacitors
1	SN7492N	TTL Divide by 12 (8 MHz output	1	0.1	disk capacitor
		only)	18	0.01	disk capacitor
1	SN7493N	TTL Binary counter (6 and 12	2	100 pF	disk capacitor
		MHz outputs only)	1	2000 μF	disk capacitor
4	SN74192N	TTL programmable up-down de-	1	1 μF	15Vdc electrolytic
		cade counters	3	250 μF	15 Vdc electrolytic
1	SN7473N	TTL dual J-K flip-flop	2	100 μF	15 Vdc electrolytic
1	SN7474N	TTL Type D flip-flop	3	30 μF	15 Vdc electrolytic
2	SN7404N	TTL hex inverters	1	10 μF	15 Vdc electrolytic
1	SN7430N	TTL eight-input *nand* gate	1	47Ω	15 Vdc electrolytic
2	SN7410N	TTL triple 3-input *nand* gate	1	68Ω	1W resistor 10%
1	LM309	5V regulator, 250 mA. (may be	4	220Ω	1/2W resistor 10%
		replaced by an LM309K if you	1	1800	1/4W resistor
		mount it off the board)	1	560	"

1	LM309K	5V regular, 1 amp., plus finned heatsink.
1	MC1648P	MECL oscillator
1	40673	RCA dual-gate-protected FET transistor
1	MFE3002	Motorola enhancemnt N-channel FET transistor (caution – don't remove clip on wires.)
6	2N706	NPN fast switching transistors, or equiv. (sil.)
1	2N5771	PNP fast switching transistor, or equiv. (sil.)
1	HEP52	PNP transistor or equiv. (sil.)
1	2N3055	(or TIP3055) NPN silicon power transistor (or equiv.)
1		12-volt 10-watt zener diode
7	1N4001	silicon diodes (1 amp, 50 PIV) (Do not substitute for the two paralleled ones in the VCO unless you know what you are doing.)
23	1N914	silicon switching diodes (or equiv.)
1		LED indicator diode – red (LOCK light) (optional)
1		12VDC 10 mA SPDT relay (Sigma 65F1A-12dc or Calectro D1-967)
2		SPDT toggle switches (CHANNEL SELECT)
2		SP3T rotary switches (MHz SELECT)
4		3P10T rotary switches (100 kHz and 10 kHz)
2		SPDT rotary switches (0-5 kHz)
1		AT-cut 10 MHz series-resonant HC-18/U crystal
1		Coil form, 7/32" dia., 15/32" long, ferrite slug (Calectro or equiv) (Coil is 3-6 μHy, adj.)

4	1K	"
1	1 meg	"
1	12K	"
1	27K	"
3	10K	"
1	1.5K	"
2	4.7K	"
2	2.2K	"
1	15K	"
1	470	"
2	100K	"
14	330	"
1	7"x9"	p.c. board
300		MOLEX Soldercon pins, used for IC sockets
1-4	PAX-1	International Crystal power amplifiers, one for each doubler or tripler
1-4	33K	1/2 resistors, one for each PAX-1 used

Miscellaneous

Chassis, case, power switch, connectors, coax, #32 enamel wire for coil, knobs, power supply providing 12V at 750 mA (or auto or motorcycle battery) – must be well filtered.

Part II-Frequency Meters and Counters

300 MHz Frequency Scaler

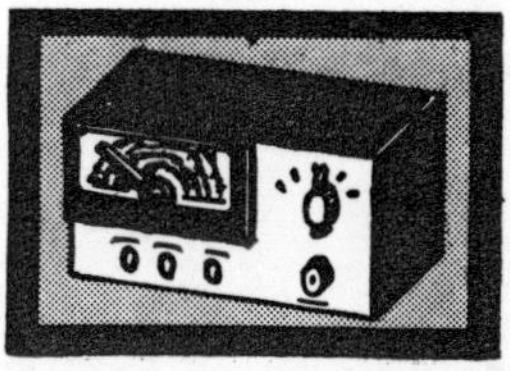

Peter A. Stark

For those who have a low-frequency counter and just need a scaler to extend the range, here is a simple design that can be built for about $20.

The heart of the unit is a Fairchild u6B95H9059X divide-by-10 scaler IC, which costs about $16 in unit quantities; we call it a 95H90 for short, and it is IC1 in Fig. 1, the schematic diagram. The output of this IC is fed into a 2N5771 PNP transistor, which amplifies the output and provides a voltage change. If you have a counter that will go up to about 15 MHz, then this part alone will extend your range to 150 MHz. But if you have a slower counter, you will need IC2 as well. This is an inexpensive SN7490N TTL decade counter, which does a second division by 10. Both ICs together will take a 150 MHz signal and divide it down to 1.5 MHz.

The ICs need a voltage between about +5 and +5.3V. While you could build a power supply, that only increases the cost. The easiest way to provide power is with four D-cells or a 6V lantern battery, which provides +6V. The 1N4001 diode, or another silicon power diode, drops this to about +5.3V. The scaler needs about 150 mA, so a set of four D-cells will last about 10 or 20 hr with intermittent use.

Figure 2 shows a printed-circuit board layout that can be used to mount the scaler, and Fig. 3 shows the parts placement on the board. Leave the copper border around the board, to act as the ground connection for the board. Use phono jacks or rf connectors, and bypass the +6V lead. The purpose is to let rf into the box only via the input lead.

In use, avoid overdriving the scaler. Start with a low input level, and feed in just enough rf to get a counter reading. With a 19 in. length of wire connected to the input lead, you should be able to activate the scaler with a 1 or 2W hand-held unit from a few feet away.

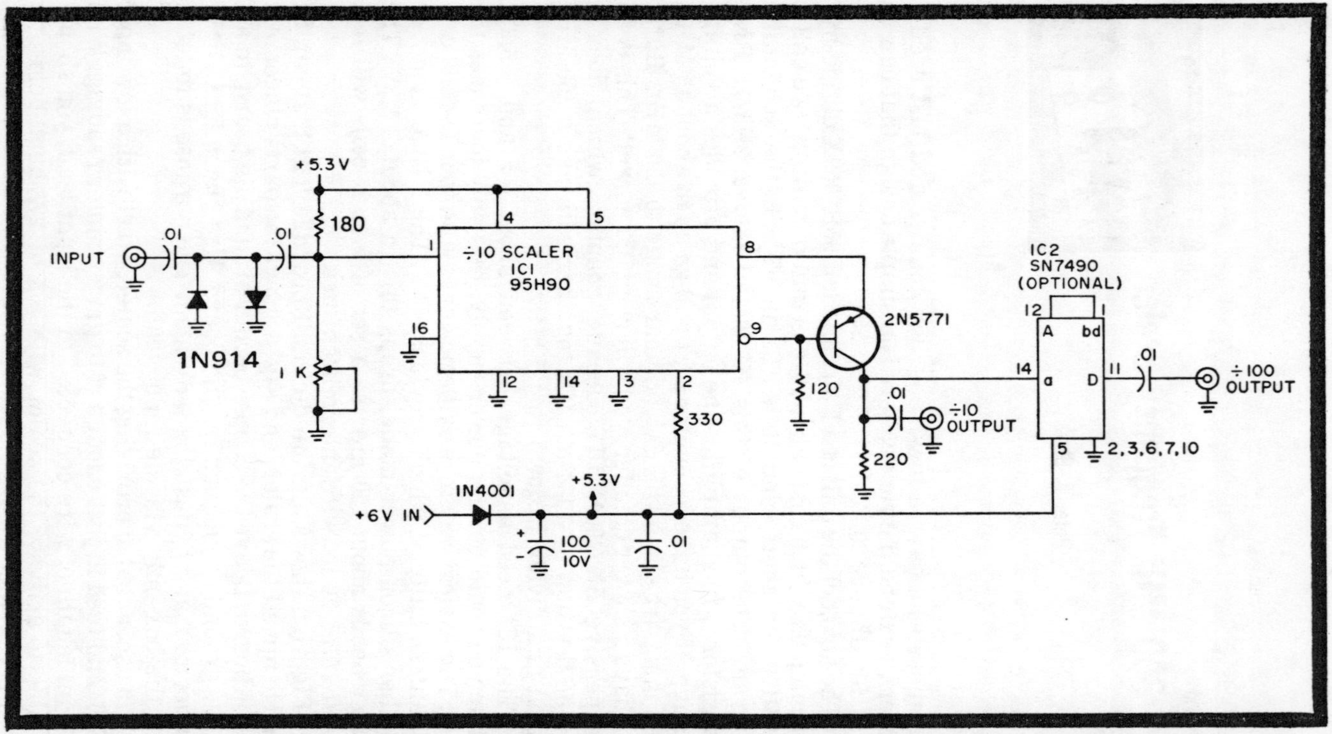

Fig. 1. The VHF scaler schematic diagram. All 0.01 capacitors are discs. All resistors are ¼W, 10 percent.

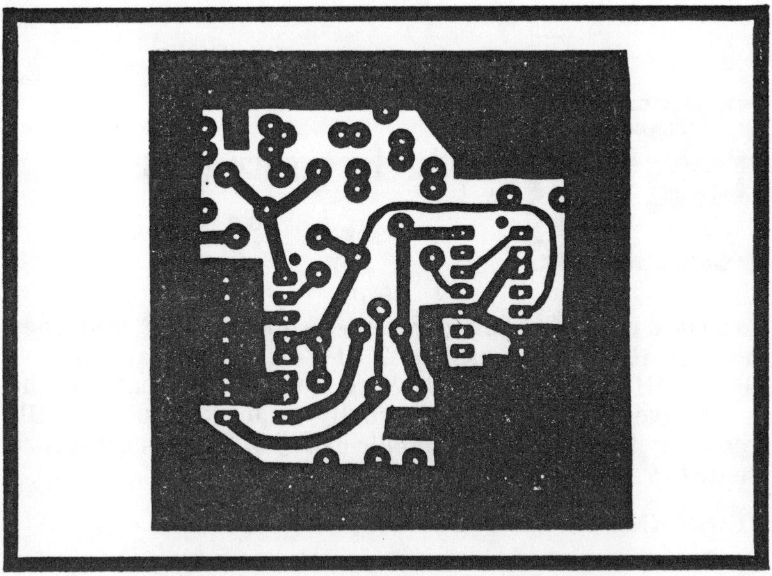

Fig. 2. Copper side of PC board (actual size).

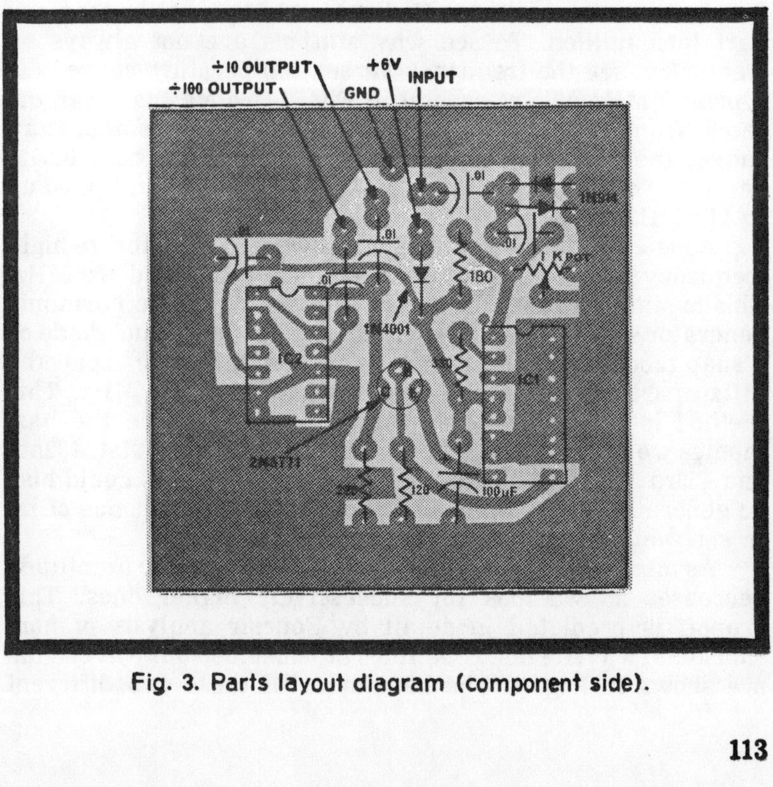

Fig. 3. Parts layout diagram (component side).

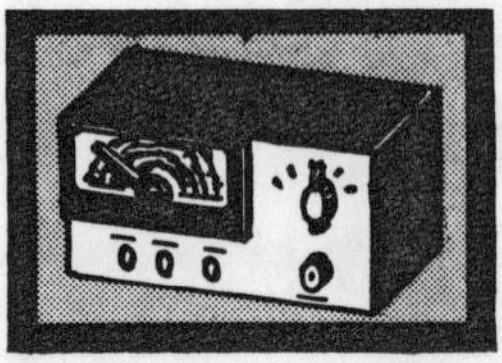

VHF Frequency Calibrator

Hank Olson

To assure oneself of being on some arbitrary VHF frequency to within 100 Hz is no easy task. If we could operate right at 144.000 MHz, it wouldn't be so hard to check; but that isn't the usual case. Rather, we are usually required to use a VHF frequency that isn't the harmonic of any of the usual standard frequency sources.

PROBLEMS IN VHF MEASUREMENTS

There are a number of ways of making frequency measurements of VHF signals, and they all have limitations. Basically, the problem is that we are trying to make a very precise measurement; 100 Hz in 144,000,000 is better than one part in a million. To see why stations are not always on frequency, see the frequency versus temperature curve of a typical 8 MHz AT cut crystal in Fig. 1. Notice that over the range from 10° C to 40° C, a reasonable room temperature range, the frequency of the crystal can vary about 0.0015 percent. This much variation is over 2 kHz, when multiplied up to 144 MHz.

A logical extension of the principles used to calibrate high-frequency receivers has been used on 144, 220, and 432 MHz. This is simply the use of a very fast switch in the harmonic generator section of a 1 MHz calibrator. With a tunnel diode or a snap diode doing the switching, useful harmonics spaced 1 MHz apart can indeed be generated through 432 MHz. This method is really the brute force approach, since the harmonics we are interested in, in this case, are the 431st, 432nd, and 433rd. Harmonics spaced at 100 kHz intervals could also be generated in the same way, but then the harmonics of interest would be the 4310th through the 4330th!

As most of us know from experience, harmonic amplitude decreases as we look for successively higher ones. This dropoff is predicted in detail by Fourier analysis of nonsinusoidal waveforms. Examples of nonsinusoidal waveforms are shown in Fig. 2. The harmonics of these two different

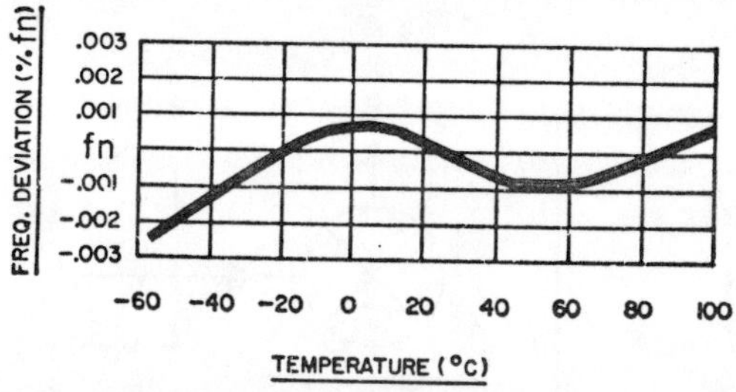

Fig. 1. Typical temperature curve of an 8 MHz AT cut crystal. Note that the crystal frequency varies up to 0.0015 percent from 10o C to 40o C, a normal room temperature range.

waveforms drop off at different rates with frequency. Note that both drop off as **1 / n** or faster (where **n** is the harmonic number). Therefore, in a 100 kHz interval calibrator for VHF use, we can expect to have **less than** 1 / 1440th of the signal for calibration at 144 MHz if the rate of falling off of the Fourier series is **1 / n**. If the falloff rate were **1 / n²**, we would have only 1/ $(1440)^2$th. A 1V 100 kHz signal, then, can theoretically produce a 1440th harmonic of about 0.5 uV, if the series falls off as $1/n^2$. Extension to 10 kHz marks will further reduce harmonic levels by a factor of between 10 and 100 (depending on whether the falloff rate is $1/n$ or $1/n^2$, respectively). To top all this off, it can be rather interesting to determine "which picket is which" in the "picket-fence" of harmonics.

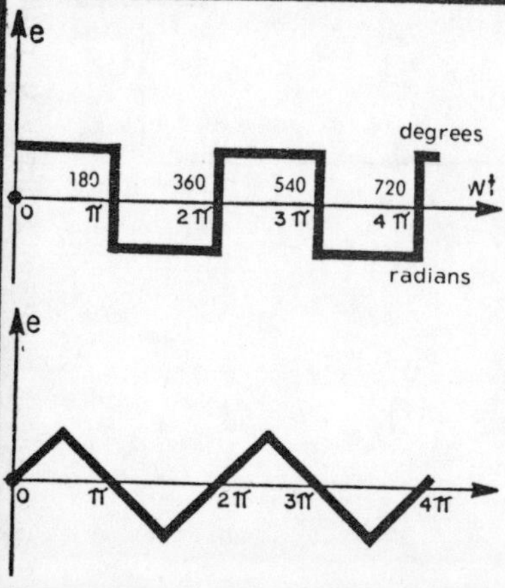

Fig. 2. Two nonsinusoidal waveforms that may be used for harmonic generation. The Fourier series of the square wave indicates that harmonics fall off at the rate of $1/n$, where **n** is the harmonic number. The harmonics of the triangular wave fall off at the rate of $1/n^2$.

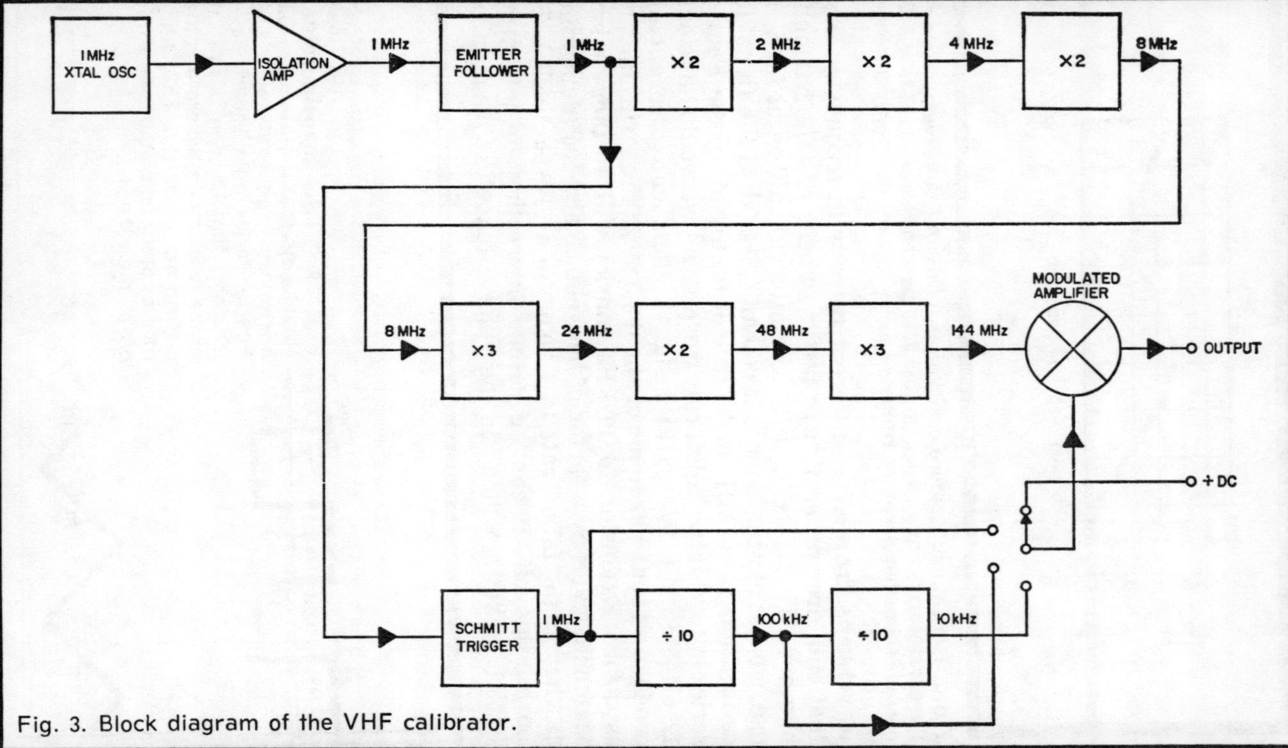

Fig. 3. Block diagram of the VHF calibrator.

The VHF calibrator presented here attempts to solve the fundamental problems of the brute force approach by applying techniques that are used in modern frequency synthesis. The circuitry is admittedly more complex, but the use of integrated circuits helps considerably to ease the construction job as well as to reduce the cost. This calibrator offers a choice of calibration modes: 144 MHz alone, 144 MHz ± 1 MHz, 144 MHz ± 100 kHz, or 144 MHz ± 10 kHz. The mode switching allows one to go from a rough 1 MHz interval frequency check to a 100 kHz interval check, and finally, to a 10 kHz interval check.

CIRCUITRY

The system is depicted in Fig. 3. Note that the 1 MHz crystal standard is both multiplied up and divided down. We produce, by means of a rather ordinary frequency multiplier chain, a clean 144 MHz CW signal that is exactly 144 times the frequency of the 1 MHz standard. This 144 MHz signal is then modulated by a rectangular wave at 1 MHz, 100 kHz or 10 kHz; and this modulation produces the desired marks. The main difference between this method and the brute force approach is that out markers now fall off in amplitude as we move away from 144 MHz (in either direction in frequency). Because we are now only interested in harmonics of the modulation frequency that are of relatively low order, the rectangular wave does not have to have a nanosecond rise or fall time. The "modulation" is not of the linear sort usually encountered, since the rectangular wave essentially turns the signal off and on.

The circuit diagram is shown in Fig. 4. Note the use of digital integrated circuits. The internal circuitry of the individual ICs isn't shown since it would make Fig. 4 vastly more complex. The Motorola HEP line of semiconductors is used for the most part, except for the four dual J-K flip-flops. These J-K flip-flops are wired to divide 1 MHz by two decades. The MC790P flip-flops (Motorola) are members of a logic family called RTL (resistor-transistor logic), which is inherently slower than MECL (Motorola emitter-coupled logic), to which the HEP digital integrated circuits belong.

If you wish to use HEP 558 J-K flip-flops to replace the MC790Ps, the circuit changes of Fig. 6 should be used. Since the HEP digital IC's are designed for +6V, a much simpler power supply and regulator are used.

The crystal oscillator sections, in both versions, use an FET as a Miller oscillator. The Miller oscillator was used here

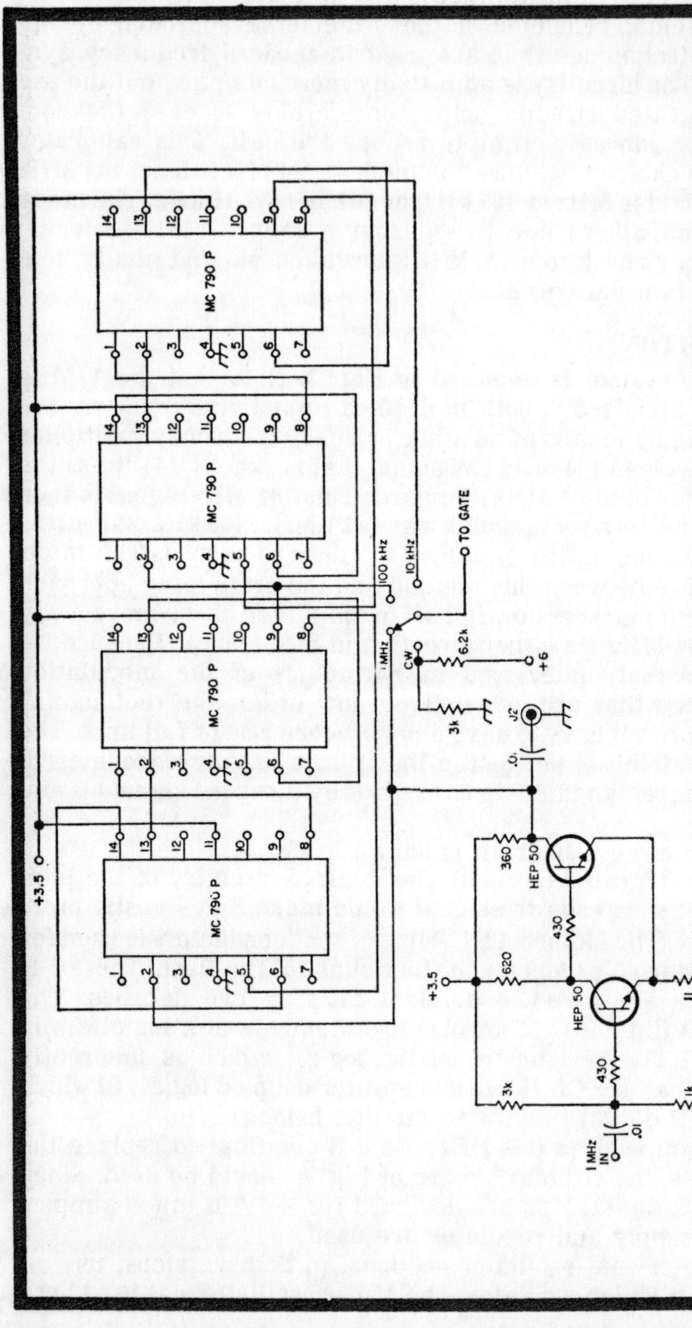

Fig. 4. Divider section uses integrated circuits to generate 10 kHz and 100 kHz marker signals from a 1 MHz crystal.

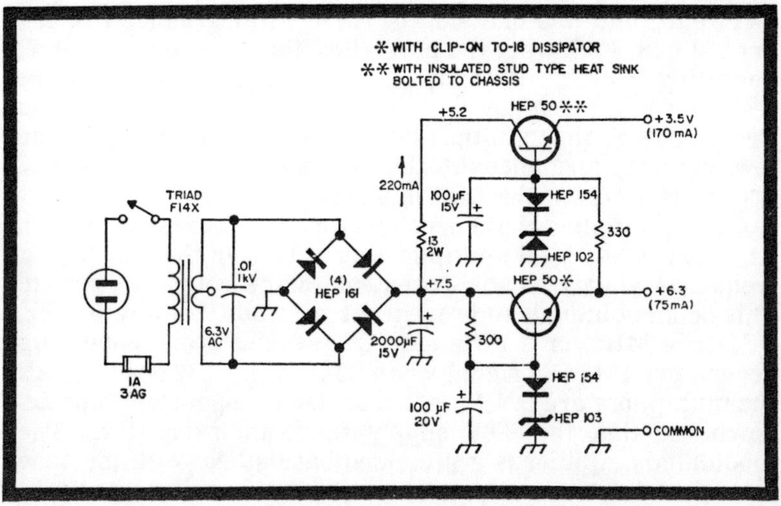

Fig. 5. Power supply section for use with the integrated-circuit divider of Fig. 4.

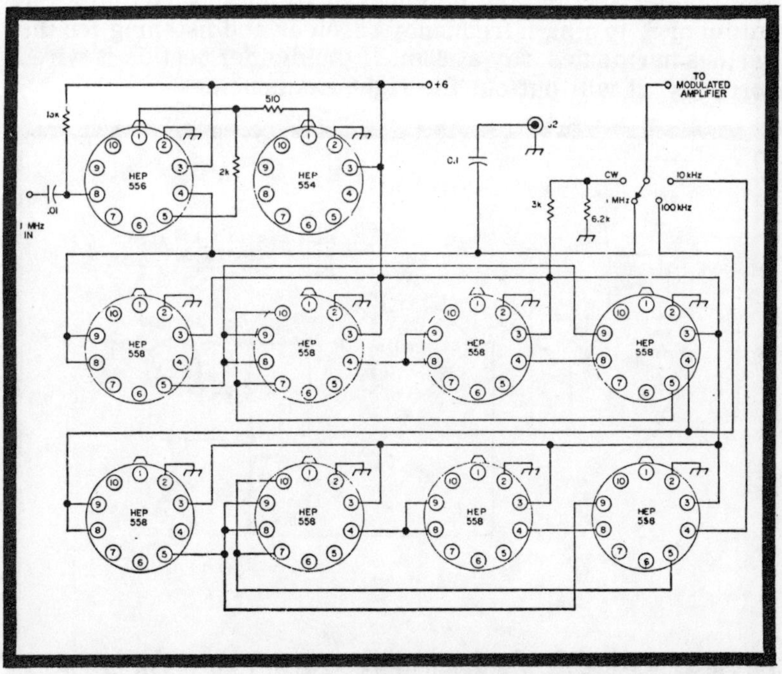

Fig. 6. Divider circuitry using HEP 558 J-K flip-flops to replace the MC790P RTL ICs, which ae inherently slower.

because the DC9AJ crystal (1 MHz) was designed for that type of circuit, and has one side of the crystal grounded to the crystal can. Following the crystal oscillator is another FET, operating as a Class A isolation stage. The isolation amplifier feeds an emitter follower, which, in turn, drives both the "count down" and "multiply up" portions of the circuitry. The low output impedance of the emitter follower is needed primarily to drive the first multiplier.

The multiplier chain is conventional in its design (x2, x2, x2, x3, x2, x3—a total multiplication of 144). In the four lowest frequency stages, double-tuned interstage coupling is used. This double tuning is to prevent the possibility of any 1 MHz, 2 MHz, 4 MHz, or 8 MHz side frequencies from appearing around our 144 MHz signal whan S2 is in the CW position. All the multipliers are PNP mesa transistors, operated "upside-down" so that the +6V supply feeds their emitters. The modulated amplifier is a grounded-base stage, with the base as the modulation control element.

Tuning of the multiplier section is easily accomplished with a grid-dip meter used as an absorption frequency meter. The divider section can be checked by loosely coupling the output of S2 to a high-frequency receiver and listening for the various harmonics, say at 80m. If the divider section is wired correctly, it will put out the right frequencies.

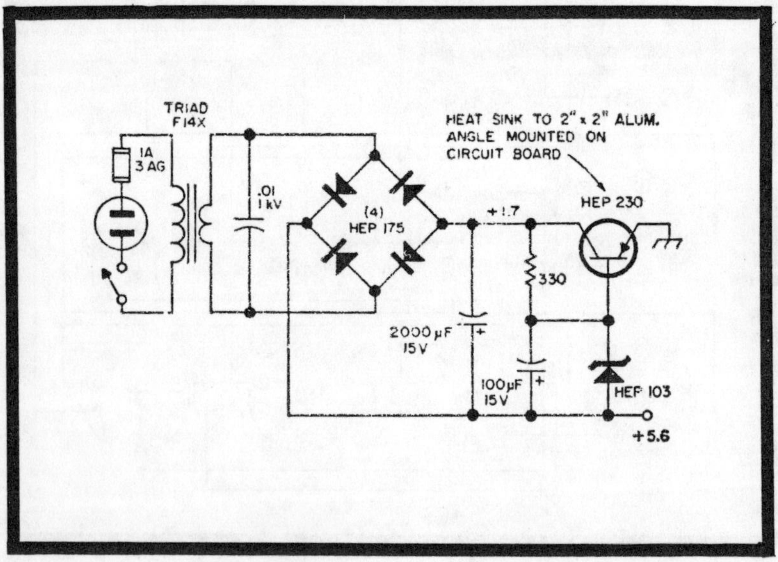

Fig. 7. Power supply for the HEP integrated-circuit divider of Fig. 6.

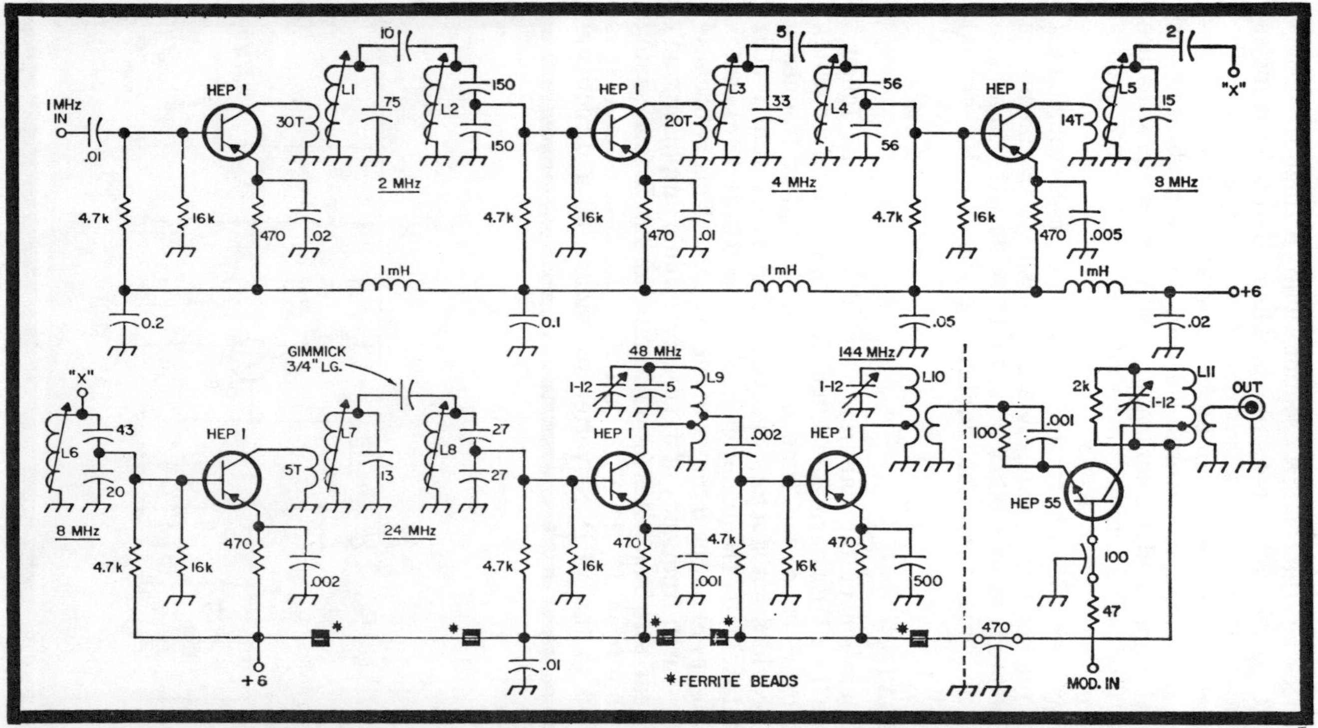

Fig. 8. The x144 frequency multiplier. A crystal-controlled input at 1 MHz provides an output at 144 MHz. Coil information for this circuit is given in Table 1.

Checking the divider section with a high-frequency receiver points out a potential problem. If the frequencies generated by the divider section are allowed to get into the receiver that is used as an i-f for your VHF converter, confusion will reign. The overall shielding of the calibrator, the general supply lead decoupling, and the VHF bandpass nature of the modulated amplifier are adequate to prevent such a problem in the units shown.

However, 1 MHz harmonics can be **purposely** coupled out (from the Schmitt Trigger) via J2. These 1 MHz harmonics are used to beat with WWV on 5, 10, 15, or 20 MHz (in a high-frequency receiver), for calibrating the 1 MHz crystal oscillator.

OPERATION

Operation of the calibrator would then be as follows. Couple the 1 MHz harmonic output (J2) to the receiver with a small (5 pF) capacitor. Tune in WWV on the highest frequency that provides satisfactory reception. Adjust C1 (the 1 MHz crystal oscillator frequency control) for zero beat. Zero beat is best observed on the S-meter of the receiver. This is because the low-cutoff frequency of the receiver audio amplifier won't pass near-zero beat notes for aural monitoring. Disconnecting the cable from J2, the VHF output of the calibrator (J3) is coupled to the VHF converter by means of a directional

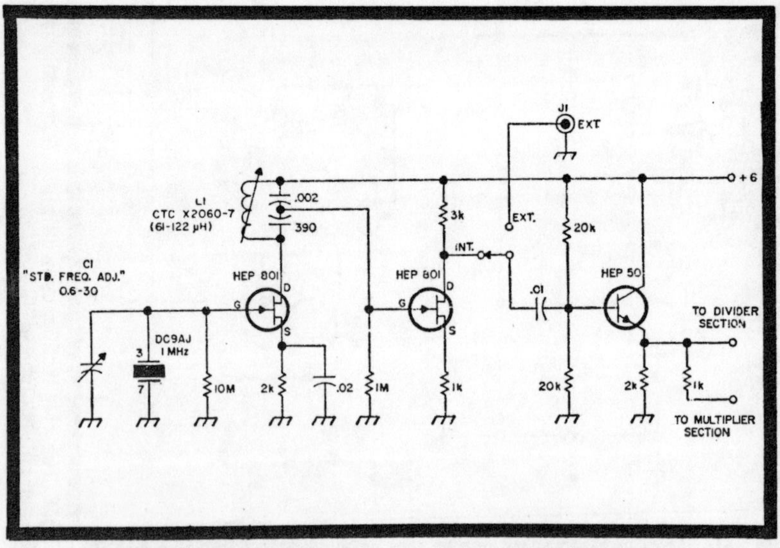

Fig. 9. The 1 MHz oscillator-buffer section.

Table 1
L1 = CTC (Cambion Thermionic Corporation) S2060-7 with 30 turns No. 28 on primary winding.
L2 = CTC X2060-7
L3 = CTC X2060-6 with 20 turns No. 28 on primary winding.
L4 = CTC X2060-6
L5 = CTC X2060-5 with 14 turns No. 28 on primary winding.
L6 = CTC X2060-5
L7 = CTC X2060-1 with 5 turns No. 28 on primary winding.
L8 = CTC X2060-1
L9 = 10 turns Airdux 416, collector tap at 2½ turns, base tap at 3 turns.
L10 = 7½ turns No. 12, ¼" inside diameter. Collector tap at 3 turns. Secondary is 2 turns No. 20 solid insulated hookup wire.
L11 = 7 turns No. 12, ¼" inside diameter. Collector tap at 2 turns. Secondary is 1½ turns No. 20 solid insulated hookup wire.

coupler and attenuator. The total decoupling between the calibrator and receiver should be about 50 dB. A temporary expedient for coupling the receiver to the calibrator may be used: A 6 to 12 in. piece of wire is simply connected to J3 to radiate the calibrator output into a nearby antenna.

Switch S2 is first put in the **CW** position and 144 MHz found on the receiver. Then the switch is set to **1 MHz**; and 144, 145, 146, 147, or 148 MHz is found (whichever is closest to the

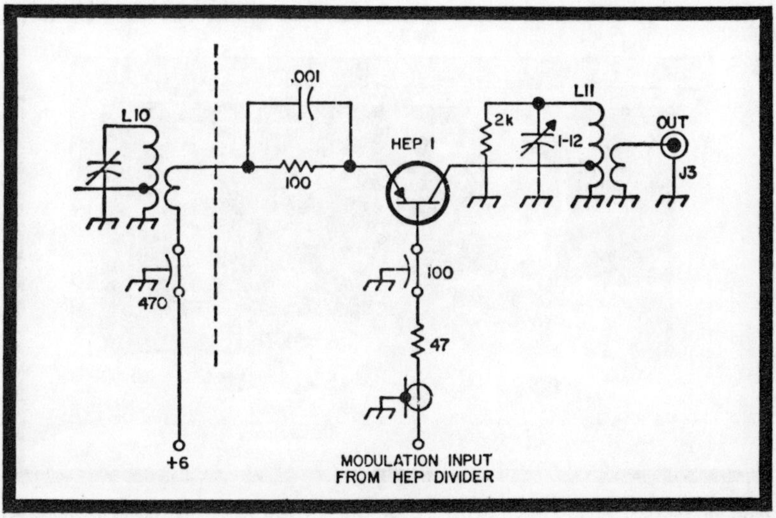

Fig. 10. Modulated amplifier for use with the HEP divider section. This circuit replaces the last HEP 55 stage in Fig. 9, when HEP ICs are used.

desired operating frequency). Then we switch to **100 kHz**, and finally to **10 kHz**, selectively pinning down our frequency.

If desired, another decade could be added to the count-down circuits to give 1 kHz intervals. Also, another tripler could be added to the multiplier chain, making the calibrator useful at 432 MHz.

Another intriguing possibility is the use of WWVB (60 kHz) or WWVL (20 kHz) as a calibration signal. By using a divide-by-5 circuit on the 100 kHz output of the first decade divider, a 20 kHz signal for comparison with WWVL is produced. By simply putting this 20 kHz rectangular wave (which is rich in third harmonic power) into a 60 kHz tuned amplifier, a 60 kHz signal is produced for WWVB comparison.

You might ask why we didn't divide 100 kHz by 5, and then 2, to obtain 10 kHz, allowing a 20 kHz pick-off after the divide-by-5 section. That was not done because it produces a **symmetrical** 10 kHz square wave for calibrator use. This type of waveform has very small even-harmonic power.

CONSTRUCTION

Construction of both units was in modular form, with the individual modules enclosed in a 8 x 17 x 3 in. aluminum chassis, which serves as a cabinet. Figures 11 and 12 show the two calibrators built by the author.

The multiplier chain assembly (which also contains the 144 MHz modulated amplifier stage) is built from copper

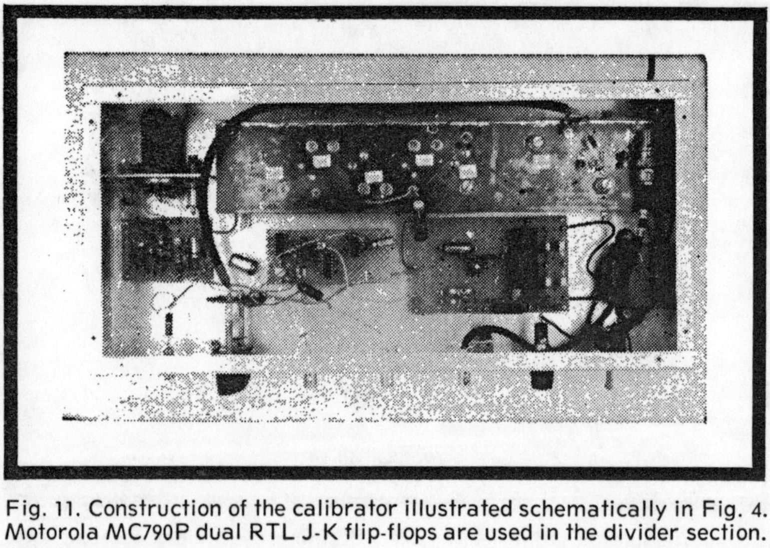

Fig. 11. Construction of the calibrator illustrated schematically in Fig. 4. Motorola MC790P dual RTL J-K flip-flops are used in the divider section.

Fig. 12. VHF calibrator built with **HEP 558 J-K** flip-flops in the divider section. This photograph shows the unit in early stages of construction, with only one decade of dividers in use. Later, four more **HEP 558's** were added to provide a second decade.

laminated board, which is used in making etched circuits. This material is easily sheared, drilled, punched, reamed, and soldered. The bottom view of one of the multiplier chains is shown in Fig. 13, and its top-plate template is shown in Fig. 14. Note in Fig. 13 that alternate multiplier stages have their transistor cans inverted; this was necessary because of the coil-mounting positions. The coils were mounted on alternating sides of the "strip," to assure stability, since there is no shielding between multipliers. There is a shield between the 48 MHz to 144 MHz tripler and the 144 MHz modulated amplifier stage, of course.

Except for the crystal oscillator, capacitor C, and inductor L1, the crystal oscillator circuitry is built on a piece of Vec-

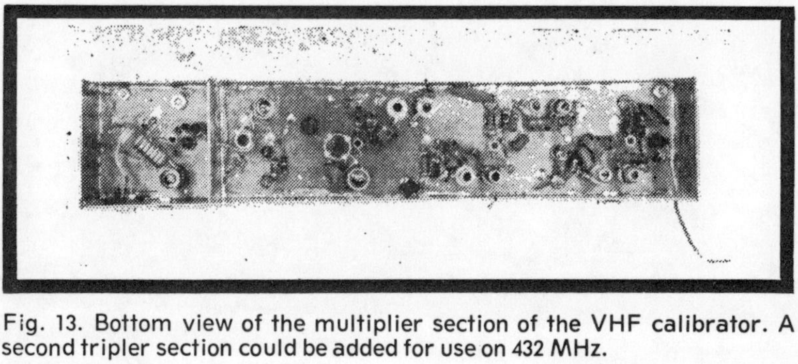

Fig. 13. Bottom view of the multiplier section of the VHF calibrator. A second tripler section could be added for use on 432 MHz.

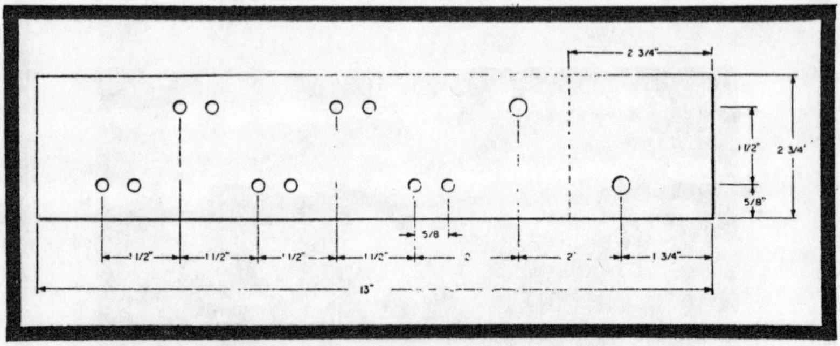

Fig. 14. Template for the top plate of the multiplier assembly.

torboard (64AA18). The crystal, C1, and L1, are mounted next to the oscillator board on a metal bracket. The metal bracket is positioned so that L1 and C1 may be adjusted through two holes in the rear of the cabinet.

The power supply is also built on Vectorboard except for the transformer and one of the regulator transistors in the dual-voltage version.

The divider units are also built on Vectorboard. Vector 64AA18 is used in the unit with the HEP ICs, with holes in the board in which to mount epoxy HEP 451 sockets. The divider unit that uses MC790P type ICs is constructed from Vector 85G24EP because the hole spacing is adaptable to the IC pin spacing. Vector pins (T28) are used for this 85G24EP board, whereas Alden 65IT terminals are used for the 64AA18 board.

Microwave Frequency Meter

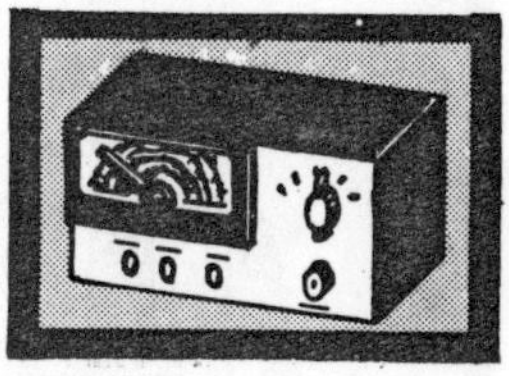

William Hoisington

This wavemeter covers the range from 1,000 MHz up through 10,000 MHz (1 to 10 GHz).

A quarter-wave coaxial cavity is used up to about 5 GHz, and from there to over 10 GHz, the three-quarter mode is used. A complete explanation of these types of operation is given.

The same type of unit can be used as a very good tuned mixer from 1 to 10 GHz.

THE COAXIAL CAVITY

The basic circuit of the coaxial cavity is shown in Fig. 1. A cylindrical outer cavity wall encloses a round rod some 4 in. long, which is the center conductor. The center conductor is grounded at one end.

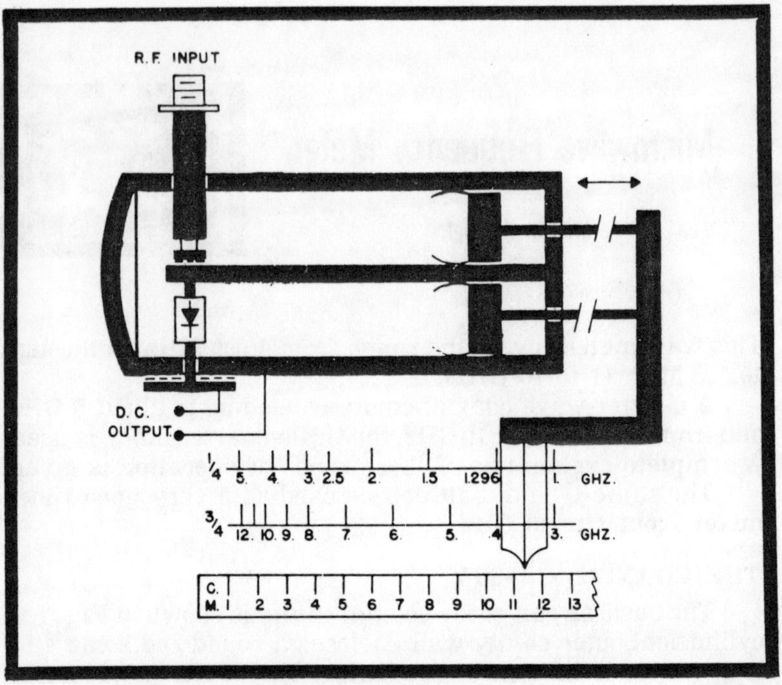

Fig. 1. Coaxial cavity, basic circuit.

The Shape of the Cavity

The exterior shape of the cavity is shown in Fig. 2, and is seen to be rectangular in cross-section, with two thin walls and two thick side walls. Believe me, this configuration was not arrived at in one day! Designing tuners for X-band, I gaily started in with sections of thin-wall round pipe, the way I'd always done on UHF. The first thing you run up against is, how do you make the diode bypass capacitor? Machine out a curved saddle piece to fit exactly over the outer wall? Possible, but too expensive. And then, how do you introduce the rf probe coupling into the cavity? Add on a "saddle" with a hole in it? These considerations and others, such as mounting (more saddles?), led to the abandonment of the pipe as a shape for microwave cavities; but not until a lot of time had been spent on the above mentioned items.

Diode Holder and Capacitor

Looking at Fig. 3, you will see the first answer arrived at; but only after weeks and weeks of making different types and shapes. The center conductor is slightly flattened and is drilled out to fit the diode prong. An 8-32 copper machine screw is drilled out to fit the other prong, slotted with a fine jeweler's saw, and compressed slightly to an inside diameter a shade less than the outside diameter of the diode prong. In this way, the copper screw will hold the diode as you insert it into the cavity. Believe me, that helps!

The second answer is also evident from Fig. 3, as the diode bypass capacity can now be made efficient at X-band. As mentioned before, you cannot buy a capacitor good for X-band use. You can make it though, as shown in Fig. 3, if the cavity body has been designed correctly for it. One of the sides of the cavity is drilled out (or machined out) just wide enough to clear the diode and its holder, which is the 8-32 copper screw. The copper capacitor plate, which is thick enough to take at least a half dozen 2-56 threads, is drilled and tapped for the 8-32 screw, and clearance is drilled in the corners for the 2-56 mounting screws. A soldering lug for the dc connection is used under one of these, and a 3 mil (3/1000 in.) sheet of fiber glass

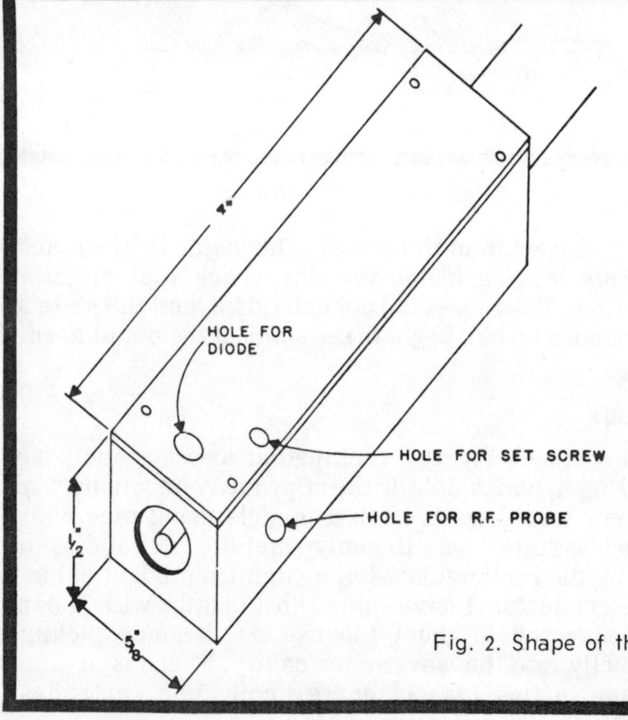

Fig. 2. Shape of the cavity.

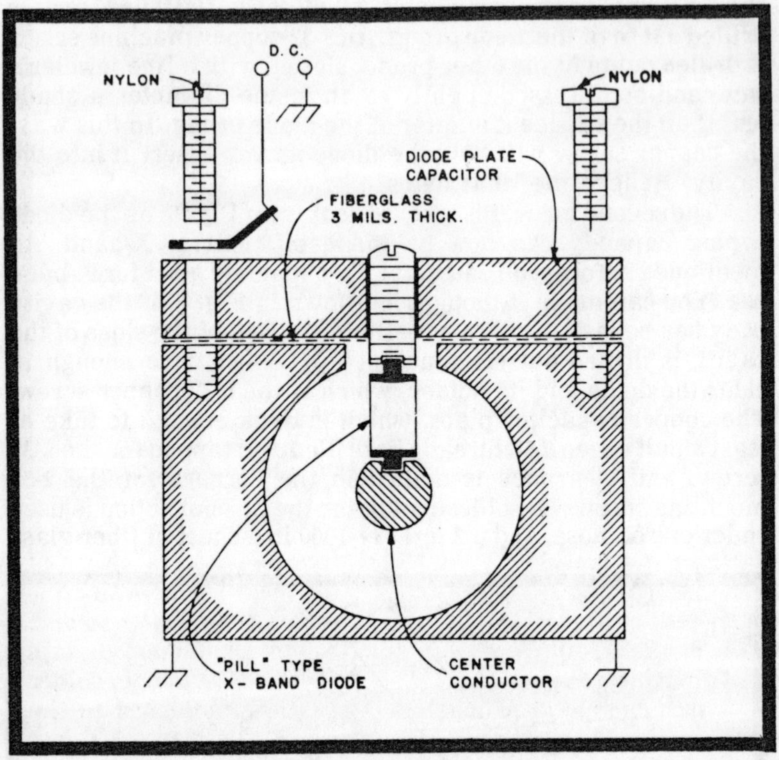

Fig. 3. Diode holder and capacitor.

cut out to fit (larger than the plate). This helps to keep metal particles from lodging inside the tiny crack that might be there if the fiber glass sheet did not extend beyond the plate all the way around. You can begin to see some of the detail needed at X-band.

The Rf Probe

Further reasons for the rectangular cross-section now show up in Fig. 4, which details the rf probe connections. This item was very troublesome in first models using pipe walls, where "more saddles" was the only solution. All saddles are eliminated by the rectangular shape. Small semirigid cable is used for the connector. I have some short lengths with X-band antennas connected to them for use as in-space pickups, feeding directly into the wavemeter cavity. There is at times an advantage in this type of energy collection (antennas), which will be taken up later.

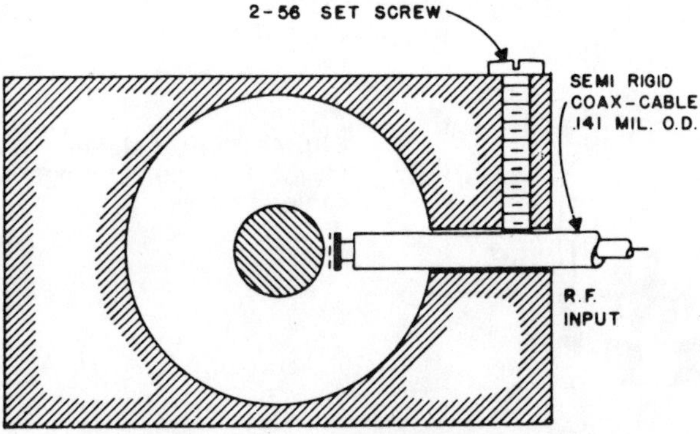

Fig. 4. Rf probe connector detail.

Figure 5 shows details of the treatment of the cavity end of the rf cable, or probe. The outer conductor is cut away for about ¼ in. and removed. About 1/16 in. of the Teflon is left, which is then removed from the center conductor. A thin copper washer (cut out of sheet copper, since the hole to solder the center conductor is quite small) is then soldered to the center conductors, as shown in Fig. 5.

Mylar tape or other good insulation is fastened to the side of this washer facing the center conductor. With this insulation in place, you can push the probe all the way in, while testing,

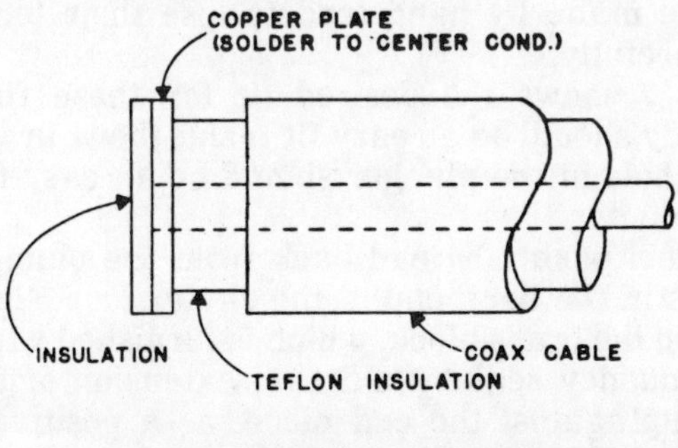

Fig. 5. Rf probe detail.

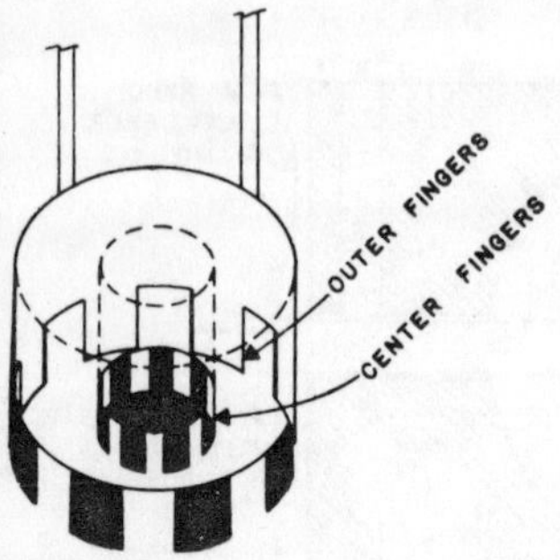

Fig. 6. Plunger details: (A) end view, (B) outer fingers, (C) center fingers.

and still not have a dead short. Different thicknesses of fiber glass sheet can also be cemented on, to make up more permanent types of fixed capacitors, of different values.

For some uses, particularly in a wavemeter, loose coupling is desired; but it must be securely locked with the setscrew, otherwise your dial calibration and frequency reading will suffer.

Plunger Fingers

Here is the most difficult item. The fingers should be made of tempered beryllium-copper, which is not easy to work with.

Figure 6 shows some details of the plunger and fingers. I assume, having been told so by "well-informed sources" (mechanical engineers), that these units should be made in a machine shop by a competent machinist. Maybe so, as the ones I have made by hand tend to lose their tension if not handled carefully.

Figure 7 shows the desired fit for these fingers. The plunger body should be an easy fit **inside** the ¼ in. cavity, and the center hole in the plunger should be an easy fit over the center conductor.

Two steel pushrods lead back from the plunger through small holes in the back end of the cavity (see Fig. 1); these terminate in the brass block, which is furnished with a pointer for the frequency scale. Maximum extension of the plunger should be up against the end piece, as a positive reference point for the dial in case of trouble after calibration. This point should be indicated on the scale as "minimum frequency," in

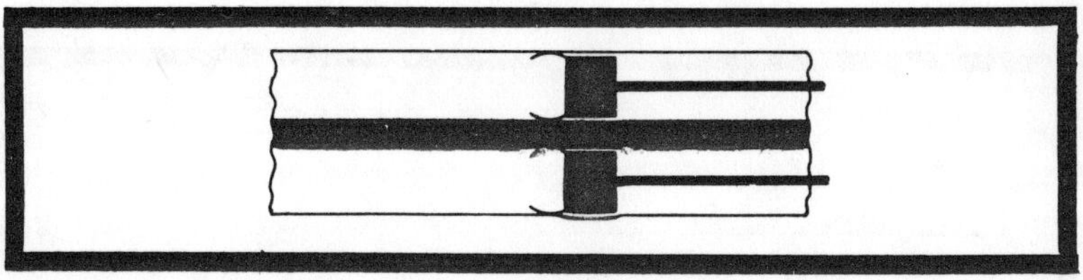

Fig. 7. Desired shape and curvatre of the plunger fingers.

order to reset the pointer if it should ever become displaced after calibration.

The Diode

At present, the diode used is an X-band "pill package," with a prong at each end, as shown in Fig. 3. These are point-contact diodes, like the famous 1N23 ceramic cartridge types of World War II fame, only a lot smaller. Referring again to Fig. 3, always make sure that the ceramic part of the diode is, as nearly as possible, in the open space between the inner and outer conductors. This space is where the rf is! It is also important to make sure that there is as much metal surface continuity as possible along the cavity wall, across the X-band capacitor insulation onto the diode capacitor plate, and from there over to the diode holder and onto the metal end of the diode.

The rf is at a maximum between the inner and outer conductors, which is an air space of a 1/16 in., and that is where the diode should be.

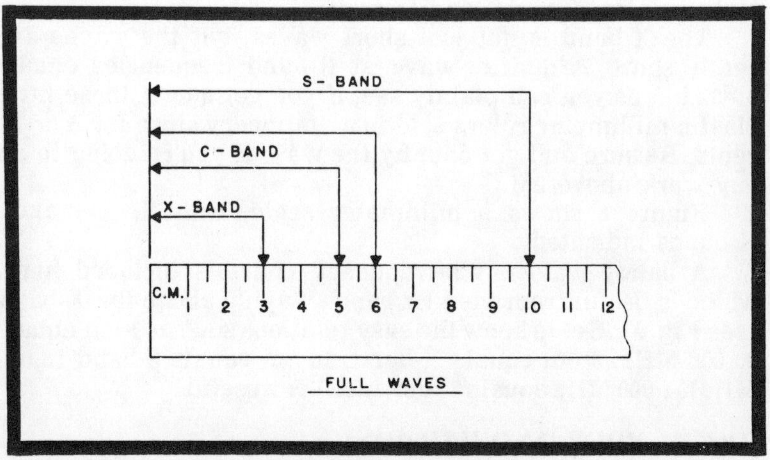

Fig. 8. Full-waves for S-, C-, and X-bands shown on millimeter scale.

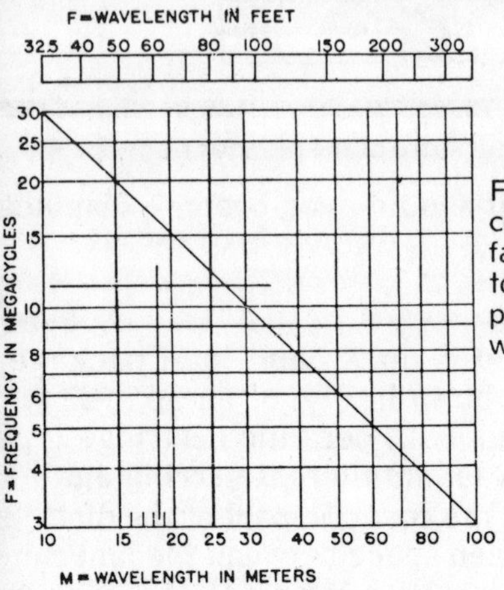

Fig. 9. Wave-length to frequency converter. Use of multiplying factors such as those at the bottom of the graph will cover any portion of the electromagnetic-wave spectrum.

The diode rf bypass capacitor, formed by the diode plate and the flat top of the cavity body, need only have a small capacity; anything over about 20 pF is sufficient. What it **must** have is the proper **lack** of inductance! The details have been covered in previous paragraphs, and if you follow those details you will find little or no rf on the **outside** of the diode capacity plate or along the dc lead from it.

The X-band is not just short waves, but the waves are really short. A quarter-wave at X-band frequencies equals 9/32 in., as you can plainly see, if you get one of those little plastic millimeter rulers sold in a stationery store for 5 or 10 cents. Be sure and get one, by the way, if you're going to do any work above 2M.

Figure 8 shows a millimeter scale, with S-, C-, and X-bands indicated.

A handy wavelength-frequency chart is included here, which is useful from the kHz range way up above the X-band (see Fig. 9). Get to know the easy relations such as 1 cm equals 30,000 MHz, 3 cm equals X-band, 10 cm equals S-band (3,000 MHz), 1,000 MHz equals 30 cm, etc. Very useful!

THE ¾ MODE AND HARMONICS

Don't worry about that word "mode." Generally, when something odd takes place in a cavity or waveguide, certain types of engineers tend to fall back on covering up. They say, "It jumped mode," or, "Spurious showed up."

Here's the straight dope. Figure 10 shows the quarter-wave mode of operation. Starting at 1 GHz you will find **one** point of maximum dc output. If the oscillator under measurement is "running hard," with lots of second and third harmonic energy content, these will be found at 2 and 3 GHz, and possibly higher. The harmonics should drop steadily in power as you go up in frequency. The diode itself may cause some of these, if hit too hard with the rf input.

Figure 11 shows the three-quarter-wave mode, which is a very "natural" type of operation. Don't forget that in an instrument of this kind you are looking for standing waves, and you want them to be of the greatest amplitude possible (within reason). So, if you tune the cavity by the plunger so that it measures about three quarter-waves (allowing for length-loading of the diode on the first quarter), you will find two peaks on the meter, due to the situation shown in Fig. 11. The

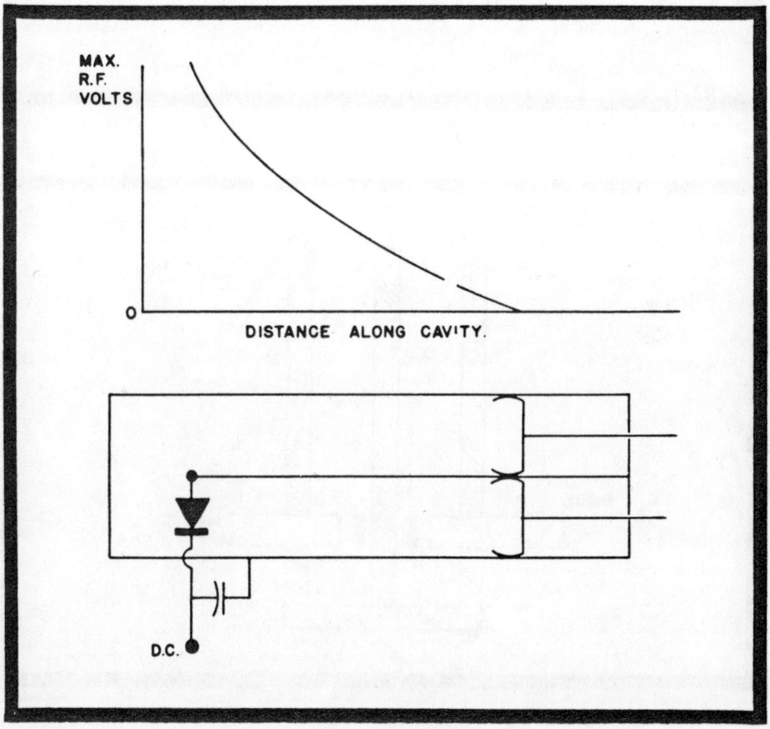

Fig. 10. Quarter-wave in cavity.

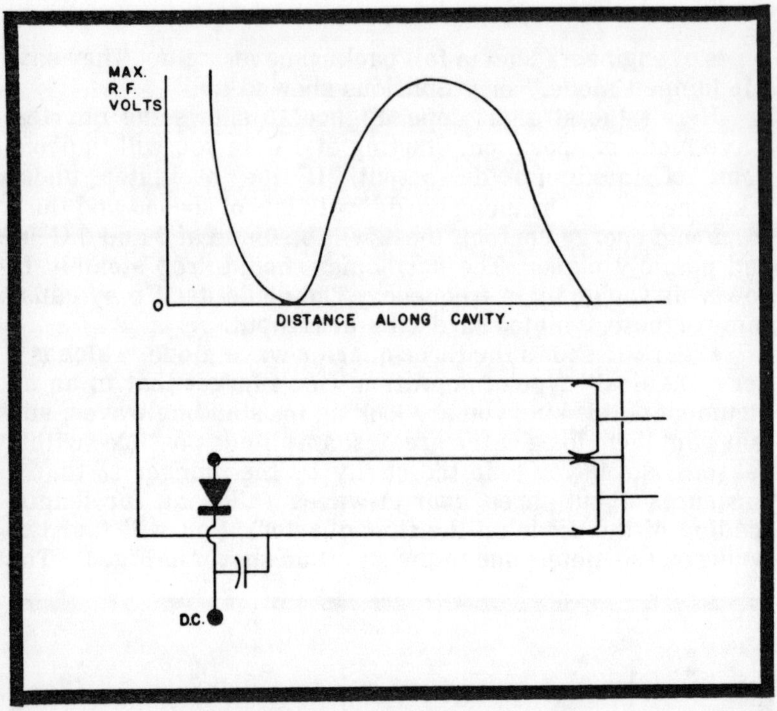

Fig. 11. Three quarter-waves in cavity.

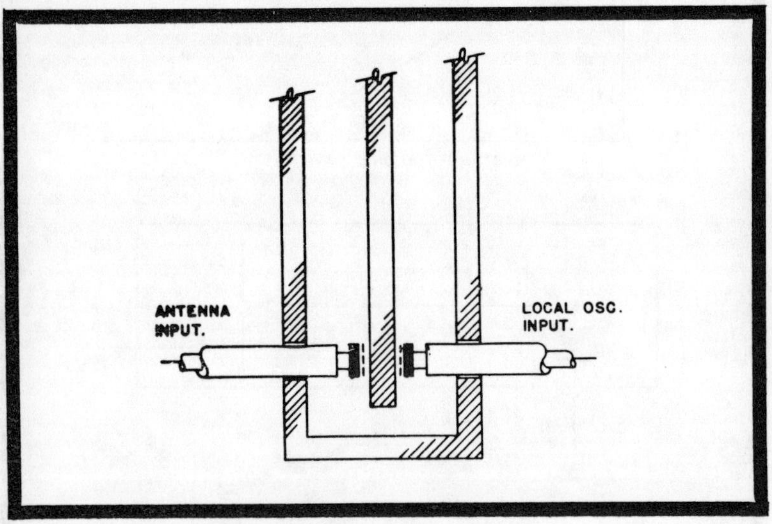

Fig. 12. Double-input detail.

higher the Q and the lower the losses along the line, the more quarter-waves can be found. For the 4 in. cavity shown, three quarter-waves at S-band frequencies are the longest that will fit.

A check on this operation is easy. Using the millimeter scale on the dial, take several readings between maximums; for example, 22, 37, 51, and 67; add the spacings together, which comes to 45 mm; divide by 3 (the number of samples); and you will find an average of 15 mm for the waves standing on the center conductor. There are 15 mm for the half-wave; 3 cm for the full wave. This is 10,000 MHz or 10 GHz (X-band).

USE AS A MICROWAVE MIXER

This same type of cavity can be used from 1 to 10 GHz as a mixer for the front end of a superhet receiver covering those frequencies. This application will only be touched on briefly here.

Looking at Fig. 12, you can see how useful it is to have **two** thick sides on the cavity, one for rf input and one for the local-oscillator input.

That about covers the details and some uses. The whole unit can be mounted on a piece of copper-clad, along with a 50 mA meter, the dial scale, and the centimeter-frequency chart. I broke down on this one and used a "regular" small microwave input connector for the rf (instead of an RCA phono jack). For connections to other units such as oscillators and multipliers, small flexible cable may be used.

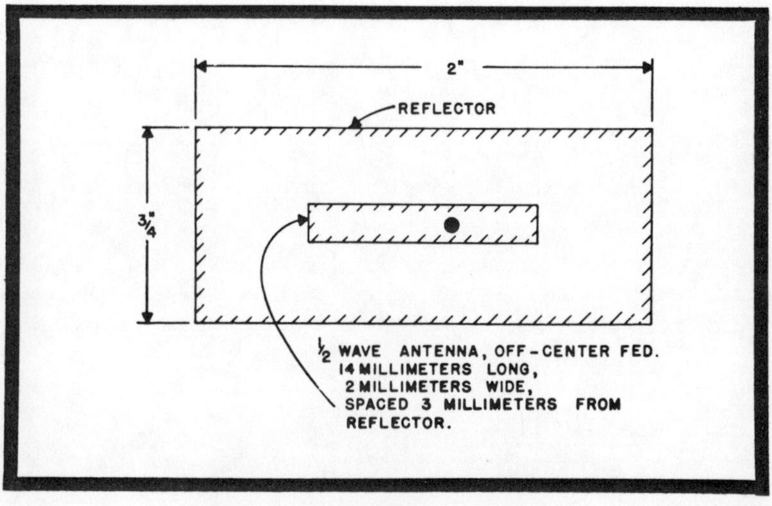

Fig. 13. Test antenna, X-band.

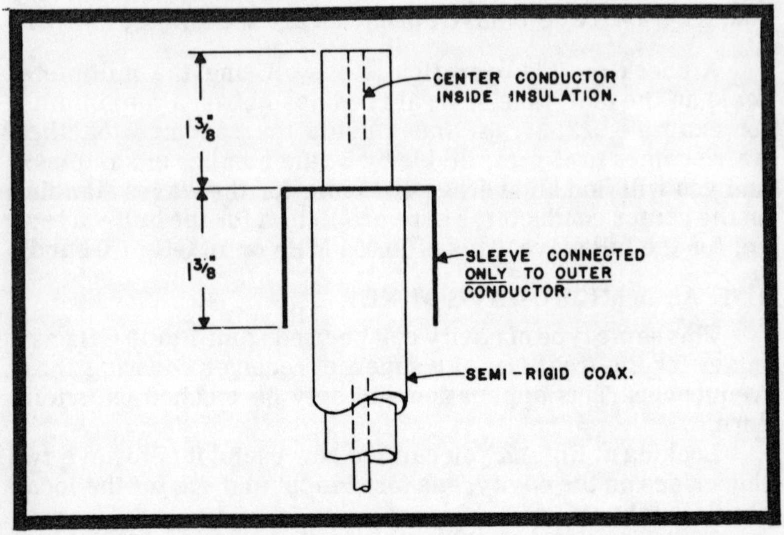

Fig. 14. Omnidirectional test antenna. Dimensions are suitable for amateur S-band, 2400 MHz.

Figure 10 shows a test antenna for X-band—not the best in the world, but good enough for a starter. With a lens in front, it really picks up signals. Figure 11 shows an S-band antenna for about 2400 MHz.

UHF-VHF Frequency Meter

Ken Sessions & Don Milbury

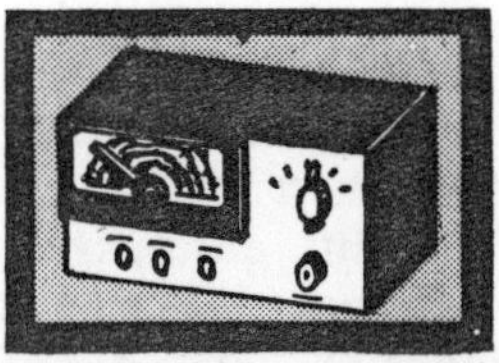

This may come as something of a shock to some, but there **is** a way to measure frequency accurately without the use of expensive equipment. With a handful of parts and a few items found commonly around the radio shop, you can build a "poor man's" frequency meter capable of tolerances that should amaze you.

This is not a substitute for a good frequency meter for commercial use, but if you have a limited number of frequencies that you want to be "dead on," this may be the answer.

A nice feature of this frequency meter is the fact that it puts to use the old surplus military communications receiver that has undoubtedly been sitting around in a dark corner of your basement under piles of old magazines and discarded dynamotors. It also uses any surplus wideband FM receiver, such as the types advertised in **73 Magazine**.

The idea is not new; persons in the two-way business will recognize it as the system that has been used for many years for the Motorola station monitor and various other common applications. Basically, it is composed of four major units: (1) a receiver converter with a calibration oscillator constructed on a high-band front-end deck from an old Motorola Sensicon A receiver chassis, (2) a monitor receiver (any 150 MHz wideband receiver—a low i-f of 455 kHz is best), (3) a WWV receiver (here is where the surplus communications receiver comes in), and (4) accessory items (hang a modulation meter on it, for example).

The block diagram of Fig. 1 shows how the individual items of equipment are interconnected to form the frequency meter. Note that although a narrowband receiver can be used, a more dependable off-frequency indication is obtainable with a wideband i-f receiver. Periodic calibration (before use) to WWV is recommended for high-accuracy measurements;

however, the unit will maintain its operating frequency to within 1 kHz (an error of 0.0006 percent) for an ambient temperature within the range of —20°C to +60°C.

PRINCIPLE OF OPERATION

The frequency converter operates on the heterodyne principle. A station frequency is monitored by heterodyning its carrier with the output frequency of the crystal oscillator and then feeding the resultant frequency of these two signals into the calibrated monitor receiver. If the beat frequency between the crystal oscillator signal and the monitored carrier is exactly equal to the frequency to which the receiver is aligned, the discriminator meter will indicate zero. If the beat frequency is lower or higher in frequency than the one to which the receiver is aligned, a direct indication of carrier frequency error in the monitored transmitter will be given on the meter.

The monitor receiver is aligned to a predetermined frequency. The specific frequency used will depend upon the spurious harmonics emitted by the channel crystals that will be required to monitor the specific carrier channels in consideration and the operating frequency to be measured.

The beat frequency fed to the control receiver may be either the sum of or difference between the channel crystal frequency and the monitored carrier frequency. Channel crystals for operation in the range from approximately 1.6 to 12.5 MHz may be used.

CALIBRATION OSCILLATOR

The calibrating oscillator consists basically of an rf amplifier stage, a mixer, and an oscillator. The calibrating crystal, shunted by a trimmer capacitor for any minor adjustment of oscillator frequency, is used for calibrating the monitor receiver.

Although the crystal is temperature controlled, a greater degree of accuracy is obtainable without the heater. The trimmer capacitor provides exact calibration of the crystal frequency at any temperature by zero beating the oscillator against the WWV signal.

The crystal heater should be used **only** when a quick check is necessary; for example, where it is desired to quickly bring the crystal to a temperature that would eventually be reached due to the heat dissipation of the equipment.

The control receiver may operate in the 145-160 MHz range; therefore, when using a 5 MHz calibration crystal, the

29th, 30th, 31st, or 32nd harmonic of the 5 MHz crystal frequency is used to calibrate the receiver to 145, 150, 155, or 160 MHz.

The selector switch operates in conjunction with the calibrating oscillator. This switch may be used to select any one of several crystals as the frequency-controlling element of the oscillator. These crystals include th 5 MHz calibration crystal and the five channel crystals.

Half of a 12AT7 tube is used as the oscillator, and the other half of the tube functions as the mixer. The carrier frequency to be monitored is picked up by the VHF antenna, amplified by the 6AK5 amplifier, and then mixed with the selected channel crystal frequency. The output of the mixer is fed to the calibrated monitor receiver, where it is determined whether the transmitted carrier is on frequency.

Assume that it is desired to monitor the output of a transmitter that is operating on 146.94 MHz, and that the control receiver has been calibrated at 150 MHz. The necessary channel crystal frequency will be the difference between 146.94 MHz and 150 MHz, or 3.06 MHz. If the transmitter is on frequency, the 146.94 MHz signal will mix with the 3.06 MHz channel crystal frequency to produce an input signal of 150 MHz at the control receiver. In this case, no indication will be given by the discriminator meter. If the monitored transmitter carrier is above or below its designated

Fig. 1. This system block diagram shows how equipment should be interconnected to make the frequency meter setup.

frequency, the input signal to the monitor receiver will be above or below 150 MHz, causing the discriminator to produce an output voltage. This voltage is fed to the meter, which is calibrated in kilohertz to give a direct reading of carrier frequency error.

When monitoring transmitters that operate in the 420-450 MHz band, the monitor must be placed so that the monitor antenna is within a few feet of the transmitter. For this application, the frequency of the stage preceding the final tripler is monitored. This is done by selecting a difference frequency crystal for the monitor that, when beat against the frequency of the transmitter stage preceding the tripler, produces the frequency at which the control receiver is tuned.

Say that it is desired to monitor the output of a transmitter operating on 443.75 MHz and that the monitor receiver is calibrated at 150 MHz. The channel crystal frequency is determined as follows:

443.70 / 3 = 147.90 (frequency actually monitored)

150—147.90=2.1 MHz (channel frequency crystal)

If the transmitter is on frequency, the frequency of the stage preceding the final tripler will mix with the 2.1 MHz channel crystal frequency to produce an input signal of 150 MHz at the monitor receiver. In this case, no indication will be given by the discriminator meter. If the monitored transmitter carrier is above or below its designated frequency, the input signal to the monitor receiver will be above or below 150 MHz, causing the discriminator to produce a positive or negative output voltage. This voltage is fed to the discriminator meter, which can be calibrated to give a direct reading of carrier frequency error.

Any error in carrier frequency indicated on the discriminator meter is an error in the frequency of the stage preceding the tripler; therefore, the error in the transmitter signal from the final amplifier will be three times as great. When using this method of monitoring, check the output (420-450 MHz) of the transmitter with a reliable wavemeter to ascertain that proper frequency multiplication is made.

CHANNEL CRYSTAL ACCURACY

Since the fundamental frequency of the channel crystals is used, any error in crystal frequency is not multiplied. Therefore, the error in monitoring a frequency by this method is very small. Crystals are held to within 0.002 percent of the specified frequency over the ambient temperature range of —30°C to +60°C. Therefore, with the previous example, the maximum frequency error of the 3.06 MHz crystal would be 3.06 x 0.0020 percent. At the frequency being monitored, the percentage error would not be discernible on the meter.

The **improvement factor**—possible percentage accuracy at the channel-crystal frequency over the percentage accuracy at the carrier frequency—is approximately the same ratio as the monitored carrier frequency over the channel crystal frequency, or 146.94 / 3.06 = 48. This is another way of stating that the channel crystal is more than 48 times better at the monitored frequency than at its fundamental frequency.

The improvement factor may be checked on any channel by the above method. It will always remain reasonably high; therefore, the possible error of the channel crystal frequency is negligible.

The front end deck of a Motorola Sensicon A receiver provides an ideal converter for this frequency meter. Figure 2 shows this assembly as a separate unit as well as in its original form installed in a Motorola Sensicon A receiver. The part numbers referred to in the modification procedures described here are the part numbers given in the Motorola manual for the Sensicon A 150 MHz receiver. The procedure is quite simple, too. Here is all you do:

1. Replace R102 (2.2M) with 3.3M, and ground the low side.
2. Remove C104.
3. Replace R103 (33K) with 470K.
4. Replace L101 with a 100K resistor.
5. Remove R112 (3.9K); complete the B+ circuit.

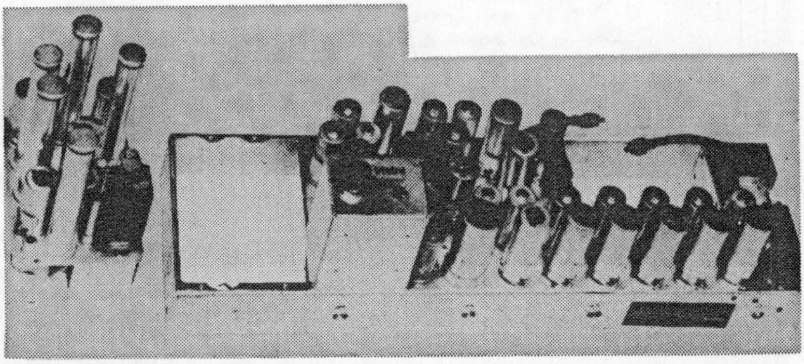

Fig. 2. The front end of a Motorola Sensicon A receiver is shown next to the receiver itself. Removal of the deck is a simple matter.

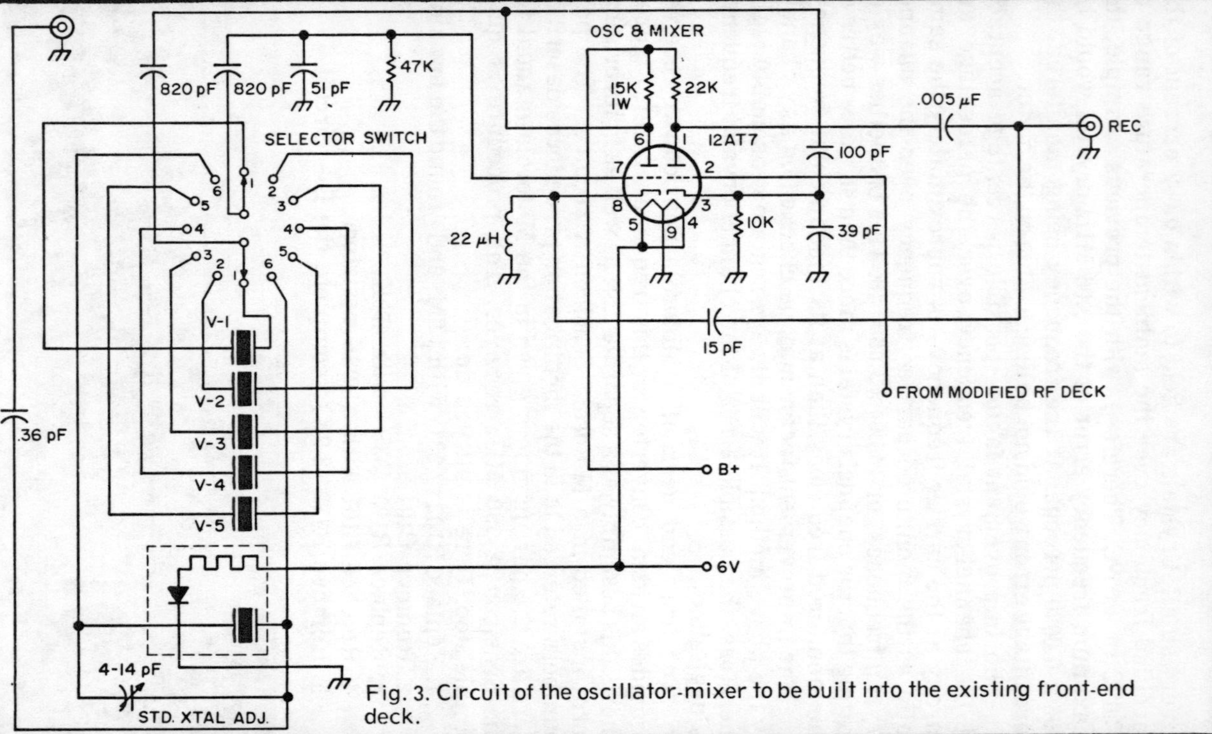

Fig. 3. Circuit of the oscillator-mixer to be built into the existing front-end deck.

6. Replace X102 with a 9-pin socket (with shield).
7. Remove the wire from pin 1 and connect it to pin 2.
8. Connect a 1M resistor from pin 2 to ground.
9. Wire the 12AT7 socket for 6V filaments. (Connect 6V to pins 4 and 5, and ground pin 9.)

This completes the modification of the rf amplifier. To construct the oscillator / mixer, remove the balance of the circuitry on the deck with the exception of the crystal socket. Then build the circuit shown in Fig. 3 around the new X102. Be sure to use silver mica capacitors in the crystal circuits. The crystals themselves are Motorola SFMT-2 (R11, 5 MHz), and they may be obtained from Sentry or International.

The Sensicon A receiver is found in a number of the older, less expensive Motorola FM units, such as the 40V, the early 41V, and all model series bearing the "AAV" and "AAD" suffixes.

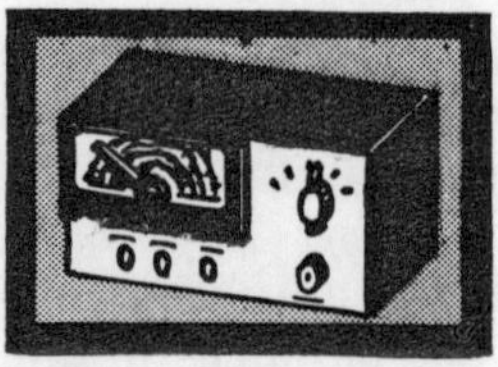

Digital Counters

Robert Suding

In building such a project as this, there are four main considerations that must be kept in mind: (1) usages, actual and possible; (2) reliability factors (bulb burnout, etc); (3) components (readout indicators, transistors, integrated circuits, resistors); and (4) complexity—number of parts, amount of wiring, etc.

DIGITAL READOUT DEVICES

Having come this far, I now wish to show you the different ways you can display a numeral; then discuss various driver circuits to light up your chosen indicator.

Numbered Light Bulbs

The simplest of all digital displays are those of Fig. 1A and B. Figure 1A is merely a piece of clear plastic, painted black on the front except where the number is. Lamps are placed behind each number and shielded from showing through to the adjacent number.

Figure 1B shows a similar method, only the light bulbs were encapsulated in plastic and little plastic letters were

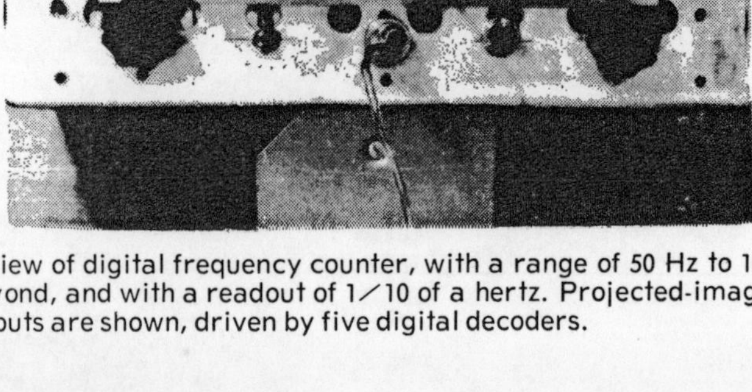

Front view of digital frequency counter, with a range of 50 Hz to 100 kHz and beyond, and with a readout of 1/10 of a hertz. Projected-image type of readouts are shown, driven by five digital decoders.

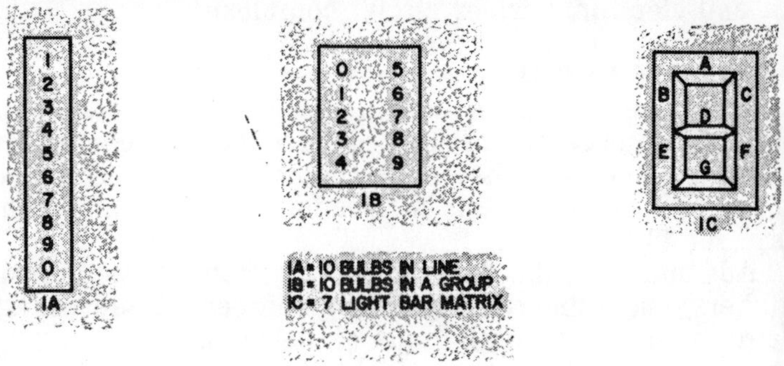

Fig. 1. (A) Ten bulbs in line, (B) 10 bulbs in a group, (C)7-bar matrix.

glued over the top of the appropriate bulb. Other arrangements—bulbs in a circle, etc.—are also a possibility of this general type.

Advantages are low cost—only the 10 bulbs and a little plastic need be bought—and simplicity. Disadvantages are relatively short bulb life; a small image, necessitating close reading; and out-of-line reading of long numbers, making the reading slower.

Light Bar Matrix

Figure 1C shows a system used by the Simpson III digital voltmeter, which presents several advantages, while adding a few disadvantages too.

Seven lights are placed so as to form 10 distinct numbers by lighting up various combinations of light bars. To get a 1, for example, bars C and F are lit; to get a 2, bars A, C, D, E, and G are lit; and so on.

These lights may take a number of forms. They may be a piece of plastic, backlighted to give a bar on the front. Another possibility is to use long lamps similar to those used in car dome lamps, only with low amperage. Even NE-2 neon bulbs may be used, along with high-voltage switching transistors.

Advantages of the light bar matrix are larger size, brighter numbers, inline reading of long numbers, and low cost in comparison with commercially available readouts. Disadvantages are difficulty in finding proper lamps, more difficult assembly as compared to simple numbered lights, brightness variation according to the number of bulbs lit, more current being drawn (due to a number of lamps being lit), and electronic driver circuit complexity.

Edge-Lighted Display

Some manufacturers have produced digital readouts that consist of 10 concentrically placed pieces of plastic, each with a number placed on it that is edge lighted to show the number. An example of this type is the readout sold by Radio Shack Corp. for $9.95.

Advantages of this readout are compactness, well formed numbers, and inline reading of long numbers. Disadvantages are dimness and a small angle of viewability.

Projected Image

This system consists of 12 light bulbs (for digits 0 to 9 and + and —), which are projected on the back of a ground glass screen. Various bulbs may be used to achieve different levels of illumination. A No. 44 bulb will give the most brilliant display, but it requires a rather high current from the switching transistor (250 mA). On the other hand, an 1819 bulb (using 40 mA) is rather easy on the transistor, but is quite dim, having a relative character brightness of 15 as opposed to 145 for the No. 44 bulb.

Top view, showing how to mount the various decoders. Section at back left is the crystal oscillator and gating circuitry. Not shown, but beneath this section, are the 24 flip-flops in the divide-by-1 million section.

Top view of digital decoder No. 2. At the top are the IC and 2 transistor gates. In the middle are the other gates. At the bottom are the drivers.

Bottom view of digital decoder No. 2.

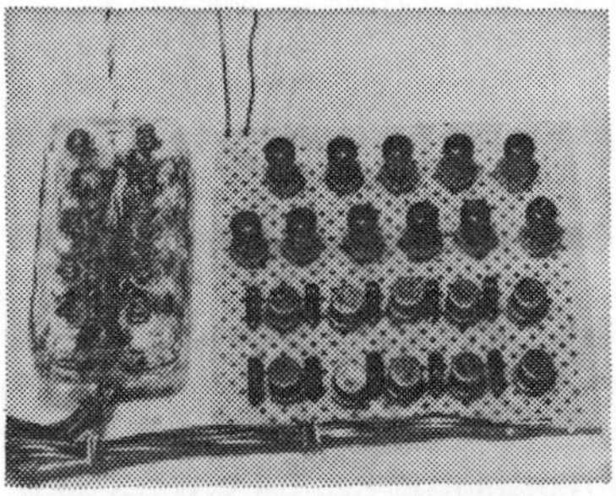

Top view of digital decoder No. 1. At the top are the five flip-flops. The second row has the 900 driver at the left, and the other five ICs are the 914s. The other transistors are the drivers. At the left is a digital display, which consists of 10 No. 49 light bulbs, encapsulated in plastic.

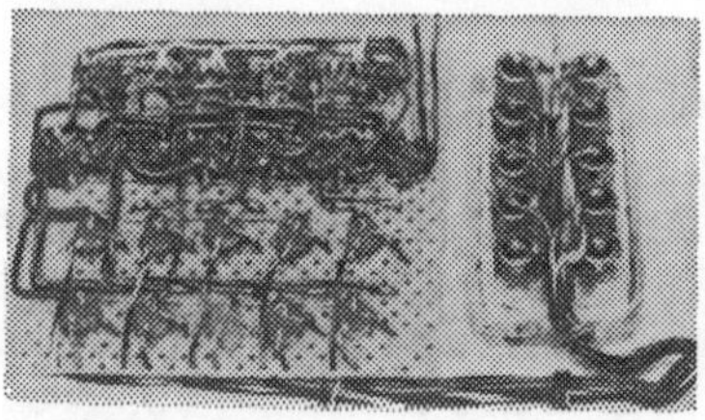

Bottom view of No. 5.

By reducing the voltage 10 percent, bulb life (3000 and 1000 hr, respectively) will be increased 5 times, giving 15,000 hours for the No. 44. Relative brightness will be cut in half. The popular No. 47 (150 mA) will perform at about half the brilliance of the No. 44.

Cost of these units varies considerably. New ones run in the neighborhood of $30 each. However, various surplus stores do have them at prices from $6 to $10, the former being about the most that I would pay. At that price, a bank of five, as I use, makes a very attractive display unit for $30.

Advantages include a nice looking image, ease of mounting, simple and cheaper driver circuits, low cost (if surplus), and inline reading of long numbers. Disadvantages are bulb burnout and bulkiness.

Nixie Tubes

These readouts are a gas-filled, cold cathode type. They are somewhat similar in basic idea to a vr (**voltage regulator**) tube or a neon tube, only they display a given number depending on which of a number of cathodes is hooked up to a negative voltage. When this happens, the gas around this element ionizes and glows. Prices on these devices vary according to size and construction. Units of interest to amateurs run from $8 to $30, the $8 one looking like a miniature tube, giving numbers 0.6 in. high. These units require special sockets running about $1 apiece. I have seen Nixie tubes on the surplus market as low as $3 apiece. For information, write to: Burroughs Corp., Electronics Components Division, Plainfield NJ 07061.

Nixies feature very long life (200,000 hr), brightness, compactness, wide-angle viewability, simple driver circuitry, and a choice of sizes. However, they require a power supply of about 200V and a special circuit for dimming.

Pixie Tubes

These are units similar to the Nixie, but instead of a relatively large number being lit up, a small number is visible through a perforated plate above the lit cathode. The advantage of these units over the more common Nixie lies in the fact that they are much cheaper, costing only $5 new and much less surplus. The main disadvantage is that the number images are so small, about 3 / 16 in. that it is very difficult to read them at a distance greater than 6 ft. Besides this, a long number would be read slower due to the individual digits being out of line.

High-Vacuum Readouts

Though similar to Nixies, these tubes project an image quite similarly to the way a crt works. For more information, write to: Industrial Electronic Engineers, Inc., 7720 Lemona Ave., Van Nuys CA 91405. This company can also give you more information on the projected-image readouts, which they make.

Advantages include long life, if the brilliance is reduced; plus the advantages of the Nixie. Also, high-vacuum readouts can be dimmed by a simple circuit. Disadvantages include a smaller image and slightly higher cost than the Nixie. Also, a HV supply of 1-3 kV is required, plus a 1.1V filament supply.

DIGITAL DRIVERS

So that you can arrive at the desired digital signal for your selected type of readout, you have three items to consider now.

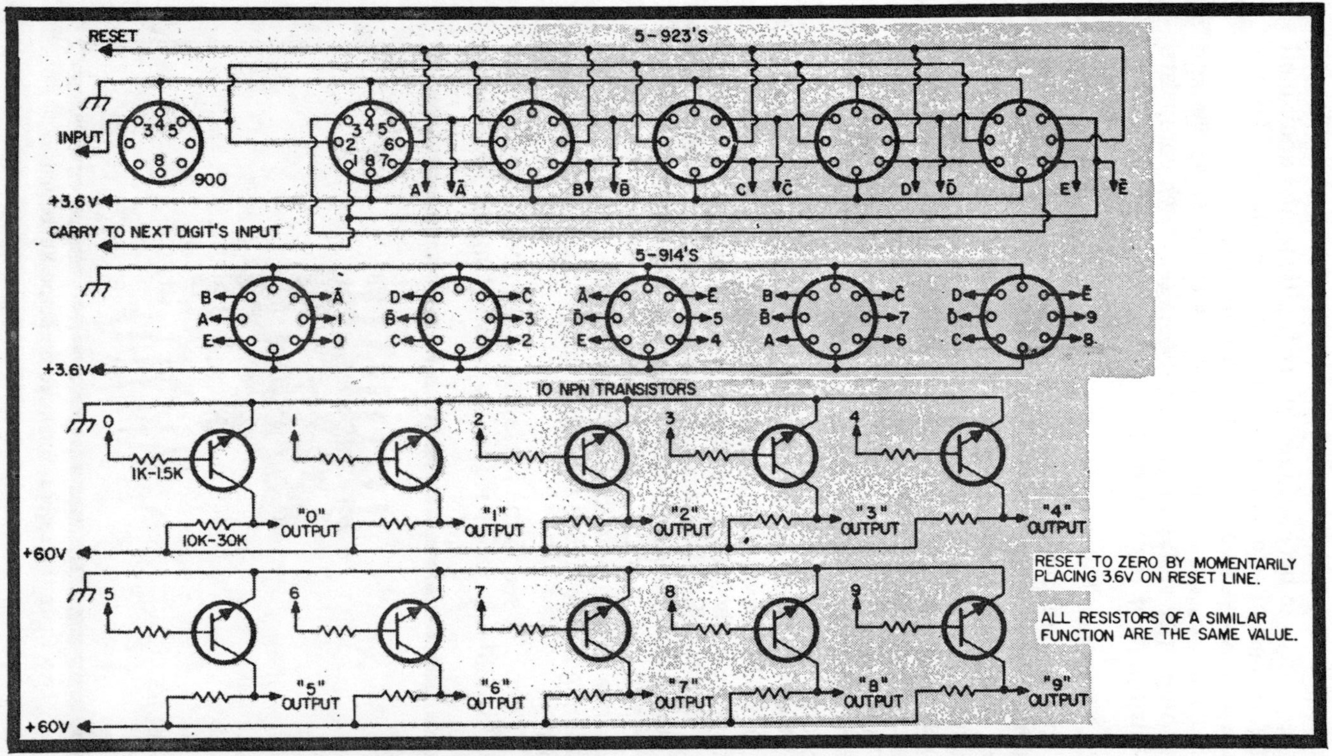

Fig. 2. Digital decoder No. 1. Reset to 0 by momentarily placing +3.6V on reset line. All resistors of a similar function are the same value.

First, you must have a set of flip-flops that will assume 10 different states, and then start over. There are a number of different ways to hook up four to six flip-flops and additional gating transistors to achieve this. In this article, I have selected two ways as being the simplest to work with.

Second, you must select the outputs of the flip-flops and steer the proper voltage to the driver stage, which comprises the third part of the whole digital decoder.

Digital Decoder No. 1

This decoder needs 31 to 41 parts: 5 J-K flip-flops (Fairchild 923 IC); 1 driver-buffer (Fairchild 900 IC); 5 dual 2-input gates (Fairchild 914 IC); 10, 1-1.5K ½W resistors: 10, 10-30K ½W resistors (only with Nixies or Pixies); 10 driver transistors (surplus NPN, or, if Nixie or Pixie is used, transistors such as Fairchild 2N3568).

The cost is $20 if you use Nixies or high-voltage Pixies and buy the resistors; $12 if you use lamps and already have the resistors.

The wiring time is about 6 hr from start to finish.

To build this circuit, a great deal of care is necessary to avoid errors. The way I wire them is to put all of the ICs in the order shown in the diagram and then wire the common pins—4, 8, and 6—of the 923s. Next, I wire the common pins of the driver and gating transistors and ICs. After this, I put in the resistors and complete the wiring except the wiring of the 914 decoding gates, which is the last wiring and the most prone to error.

As you can see in the parts list, there are a few different ways to build this decoder. If you plan on using the Nixie-Pixie

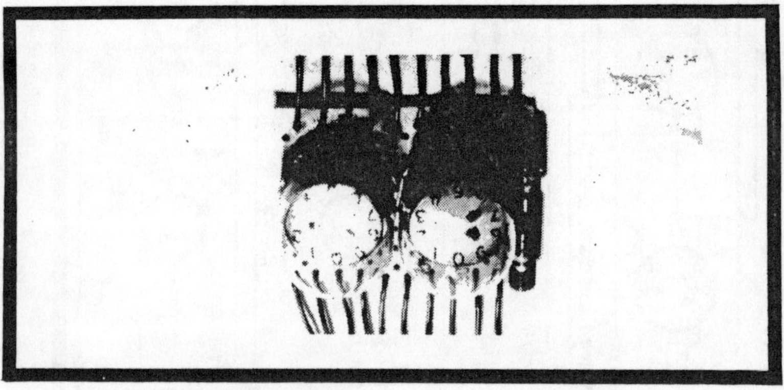

Two Pixie indicators, mounted on some keyboard.

An edge-lighted indicator, showing internal construction.

type of indicator, then you will need to use such transistors as the Fairchild 2N3568, or others with a 60V or more collector-to-emitter voltage rating. The 60V line and 10 resistors to the driver transistors are also only necessary when Nixies or Pixies are used also. Figure 3A shows how to hook up lamps to the decoder, and 3B shows how to hook up Nixies or Pixies.

Digital Decoder No. 2

This is the unit that I use in my present digital counter. It has only four ICs and uses 22 NPN computer transistors. This, while cutting down on the cost, adds to the complexity.

Parts used are as follows: four J-K flip-flops (Fairchild 923 IC), 22 NPN switching transistors (surplus computer), two 2.7K ½W resistors, five 470-680 ohm ½W resistors, ten 4.7-6.8K

½W resistors, and ten 1-2K ½W resistors. The cost is well under $10; the wiring time is about 6 hr.

As you can see, there is a definite cost advantage to decoder No. 2, if you can come by the transistors and resistors cheaply. I find the wiring of decoder No. 2 easier, due to fewer wires in the decoder, which is the most confusing part. Should you desire to use decoder No. 2 with Nixie or Pixie readouts, then change the driver transistors to HV types, and add a 10-30K resistor from each driver's collector to a +60V source, as in decoder No. 1. Please note on the decoder No. 2 diagram (Fig. 4) that the outputs are not in numerical order. This is due to the fact that the ICs are in a **biquinary** count configuration, which means that they count to 5 twice to reach 10, and then start over.

REFINEMENTS

A simple count checker that can be used to see that everything is counting correctly is shown in Fig. 5. Before you hook up the counter to read a multidigit number, make sure that you have each digit counting correctly!

Figure 6 shows the addition of an emitter follower to my original input circuit; this considerably improves the low-frequency performance.

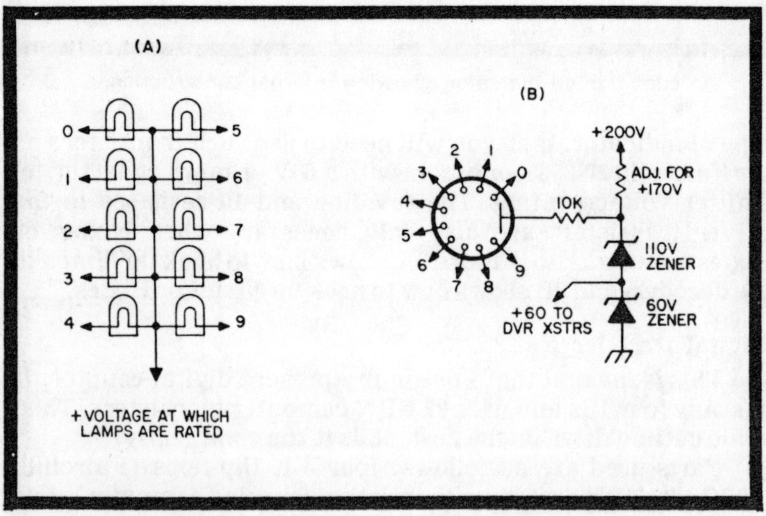

Fig. 3. (A) How to hook up lamps to the decoder. (B) How to hook up Nixie-Pixies.

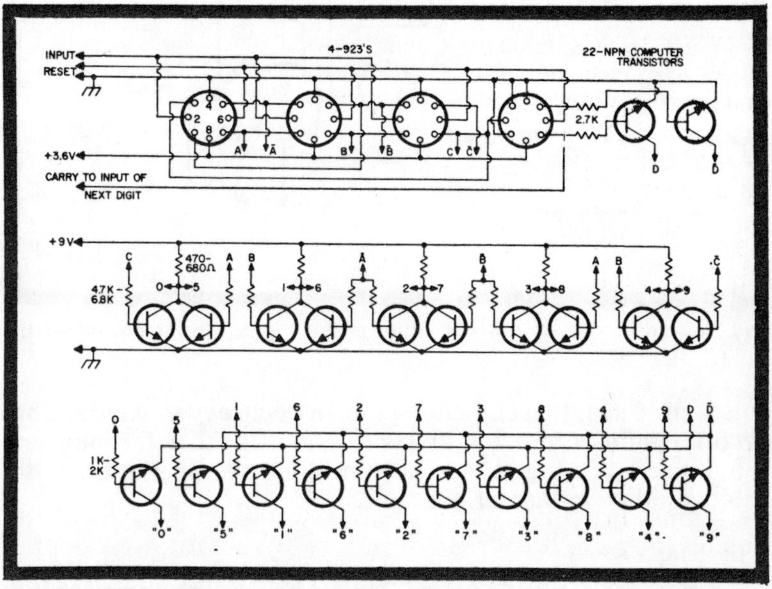

Fig. 4. Digital decoder No. 2. All resistors of a similar function are the same value. To reset to 0, place +3.6V on reset line momentarily.

Digital Frequency Divider

As many have mentioned to me regarding the binary counter, it would be a shame to depend on the 60 Hz line frequency for a timing standard for such an accurate instrument as a digital counter. Therefore, I am now using a 100 kHz crystal as the standard. Figure 7 shows that I am using a crystal circuit and a 914 monostable pulse shaper. This in turn drives twenty-four 923 flip-flops hooked up as six divide-by-10 frequency dividers. This results in a division by 1 million, giving me a final output frequency of 1 Hz in 10 sec. Using this to trigger the count controller and gate, I take a count of all

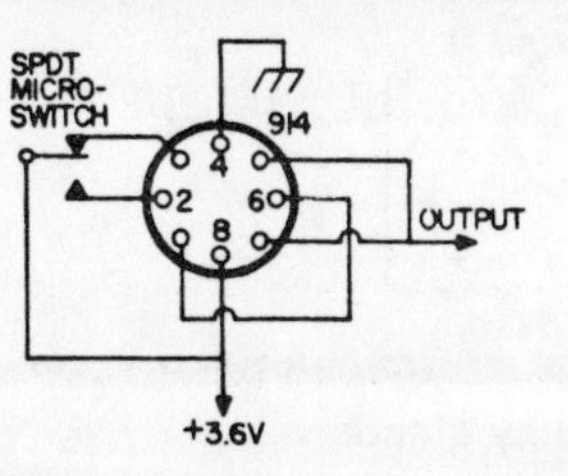

Fig. 5. A simple count checker.

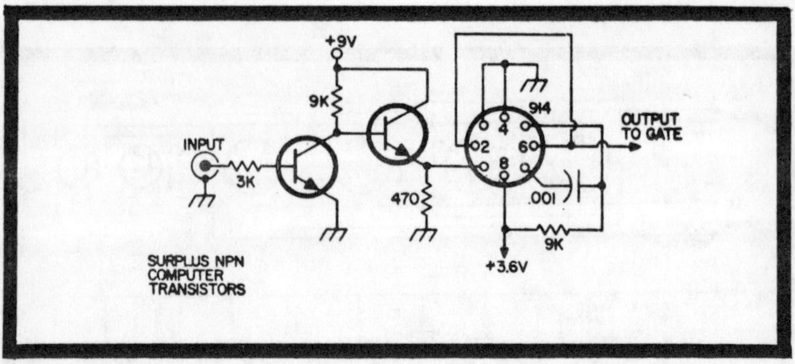

Fig. 6. Addition of an emitter follower to original input circuit of the binary counter.

pulses in 10 sec, thereby giving me an accuracy to 0.1 Hz. The decimal point is inserted as hertz. Various other frequencies are tapped off the divide-by-1 million set of flip-flops to achieve other values of counts.

Flip-Flops

A number of people have inquired about other types of ICs that might be used. Suppose that you wanted to count up higher in frequency. To do this, you would have to change the time constant of the input shaper by putting in a smaller capacitor, and then using higher frequency ICs. Say you wanted to read the output of a 7 MHz VFO. Simple—use Fairchild 926s (frequency to 8 MHz as opposed to 2 MHz for the 923). Make a frequency divider section that will give you the desired accuracy, say, 100 Hz. This will require a divide-by-1000 section, or 12 flip-flops. Now you can read out the frequency of that VFO to, say, 7034.8 Hz. Exciting? Certainly is! Accurate? Just as accurate as when you zero that 100 kHz

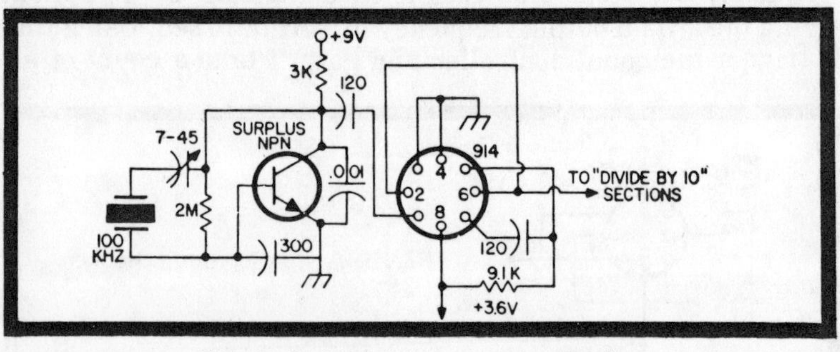

Fig. 7. Crystal frequency standard.

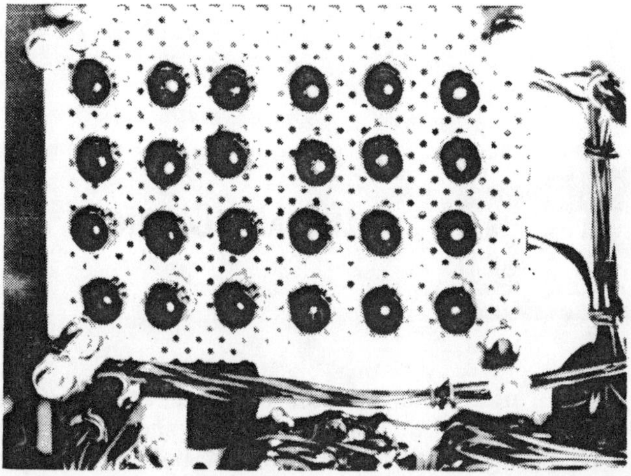

Top view of the twenty-four 923s in the divide-by-1 million section.

crystal with WWV. Of course you don't have to read out the whole number, but can read out the last three or four digits only, since you should already know the frequency to within a megahertz.

Bibliography

For those persons who are interested in the theory behind digital readouts and decoders, I would like to recommend the following free literature.

Nixie-Pixie Tubes. For bulletin 1104A, bulletin 1095, and publication No. 616E (general catalog and applications), write: Burroughs Corp., Electronic Components Div., Plainfield NJ 07061.

High-Vacuum Projected-Image Readouts. Request information from: Industrial Electronic Engineers, Inc., 7720 Lemona Ave., Van Nuys CA 91405.

Fairchild ICs. For applications brief 36 on the 9960 series, circuit notes RTL 1 through 5, App-120/2, App-118/2, and SL-218 write: Fairchild Semiconductor, 313 Fairchild Dr., Mountain View CA.

Fairchild Parts and Information. Request "Designing with Integrated Circuit Components." Write: Hyer Electronics Co., Denver Technological Center, PO Box 22227, Denver CO 80222.

General Digital Information. Write: Interstate Electronics Corp., 707 E. Vermont Ave., Anaheim CA. Request S-139A and updates.

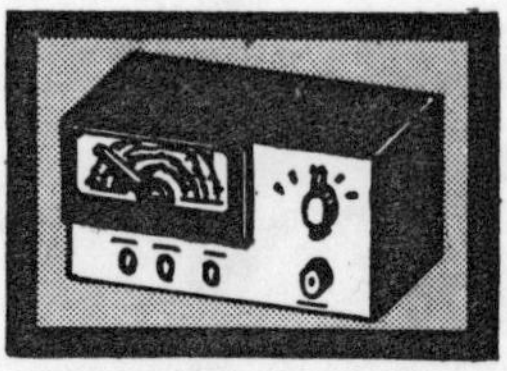

100 kHz Counter

Robert Suding

After reading numerous articles on integrated circuits and frequency counters, I decided that there should be a way to build a frequency counter that would be both simple and cheap. There was! It features 24 inexpensive ICs, 15 surplus transistors, a frequency range from 20 Hz to 100 kHz (extendable), binary frequency readout, and fully automatic operation. All you have to do is connect the input leads, turn it on, push the **count** button and read the frequency. In addition, this counter is small, takes less than a week to build, and costs less than $50.

USES

This counter has a number of standard uses, plus quite a number of special uses, which I am still studying. Some of the more obvious uses are:

(1) Use it to quickly and surely tune an audio oscillator. (I tuned a tuning fork with it to within one-ninth of a hertz.)

(2) Use it to tune teletype toroidal coils. (Build a small transistor oscillator that will use the coil and capacitor to be tuned, and then measure the frequency while substituting various capacitors.)

(3) Check the amount of shift on a fsk signal. (Measure the mark signal, then the space signal; subtracting the difference will give you the amount of shift.)

(4) Measure the frequency drift of a receiver or transmitter. (Obtain a beat note from a stable source; measure and graph the resulting change in count.)

(5) Check how far individual stations are off a net frequency. (Measure the beat note between a given station and a frequency standard.)

(6) Use it as a chronograph for gun and racing activities. (Feed in a known frequency by turning the counter on and off

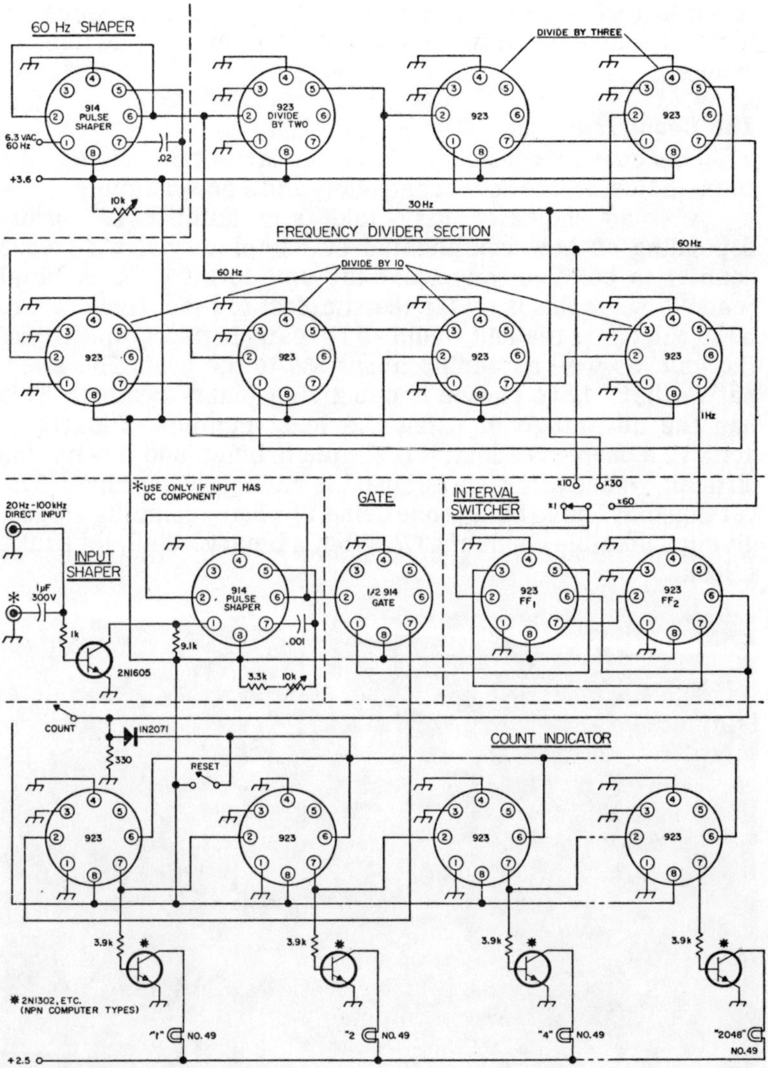

Fig. 1. Schematic diagram of the cheap and easy frequency counter. This counter uses inexpensive integrated circuits and costs less than $50 to build.

with start and stop traps and reading the counts; then convert to speed by using the following formula: $V = fd/n$, where V = speed in feet per second, f = frequency of source signal in hertz, d = distance between traps in feed, and n = number of counts.)

The Count Indicator

This counter consists of three fundamental sections: the count indicator, the count controller, and a power supply.

A count indicator may take any number of forms, depending on how complicated or complete you wish your counter to be. You may want the unit to give you decimal readout, since this is by far the simplest to read. However, to build a decimal readout would add greatly to the complexity of the unit, as well as adding about $50 to the cost. The alternative that I have chosen is one giving binary readout. This has the advantage of using the least number of parts to achieve a usable readout, it is simple to build, and it is not too difficult to read after you get used to adding the numbers. The version that I have built is one using 12 **binary** counting stages, giving a possible count of 4097. What's binary? Well, let's take a look.

Top view of the count indicator (middle), the count controller (left), and the power supply (right). The three terminals on the top of the count controller are used when the unit is used as a chronograph (for measuring bullet speeds, for example).

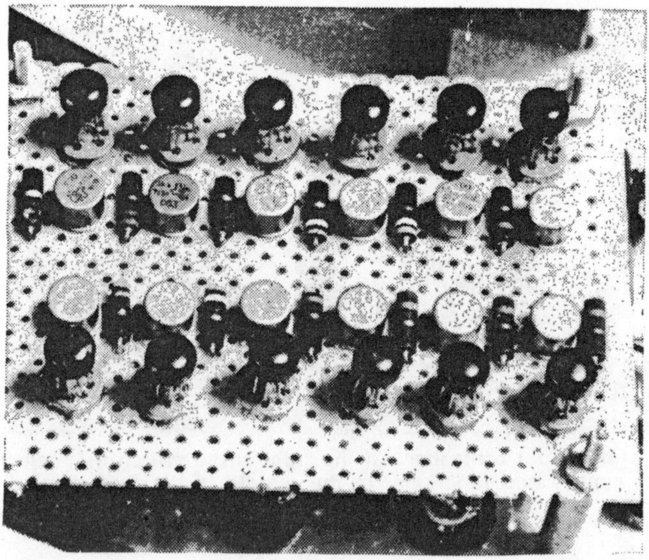

Inside view of the count indicator, showing how the 12 stages of ICs and associated transistors are mounted on punched board which measures 3 x 4¼ in.

The various stages of the count indicator are consecutively labeled in powers of 2, starting with 2^0, which has the value of 1.

At the start all bulbs are off. On the first pulse, the 1 bulb turns on; the rest stay off. The second pulse turns the 1 bulb off and turns the 2 bulb on. The third pulse turns the 1 bulb on; the 2 bulb stays lit, and the rest stay off. Adding this total gives a count of 3. The fourth pulse turns the 1 and 2 bulbs off, and turns on the 4 bulb. The process continues for 12 stages on the model I have built, for a total count possibility of 4097 (1 less than twice the value of the highest bulb in the counter). In summary, to read binary just add up the values assigned to the bulbs that are lit.

Construction of Count Indicator

If, in building this section, you put all of the parts in the same order and orientation that they have in the count indicator diagram, you will find that it can be built in just a few hours. I used miniature 8-pin integrated-circuit sockets, but you could just as easily solder them in. I bent the No. 4 pin up on all 12 sockets, and the No. 8 pin down. I then connected a wire between all of the No. 4 pins for a ground and minus lead,

Inside of the count controller. At the right is the switch that selects the various ranges and functions. In the middle is the frequency divider section and interval switcher with gate. This punched board measures 3¼ x 2 in.

and another wire to all No. 8 pins for the +3.6V line. Next, I ran a wire from each No. 1 and No. 3 pin to the No. 4 pin, so all three pins are grounded. This completed the power wiring. Next, I connected all No. 6 pins to the reset button, and completed the wiring. The resistor in the base of each of the transistors had to have its value juggled around a bit to make the bulbs glow evenly. Lowering the value makes the bulbs brighter, and raising it dims them.

In the interest of simplicity, I mounted the indicator light bulbs on the front panel inside some grommets, as you can see in the picture. Instead of using sockets, I also soldered wires directly to the bulbs, and adjusted the power supply so that there is only about 1.5V across each bulb.

The Count Controller

To put it in the simplest terms, the count controller must turn the count indicator on for 1 sec, then off again, so as to see how many hertz there were in that second of time.

The count controller consists of five sections. The main section is the one that forms the 1 sec interval. I achieved this by taking the 60 Hz ac line frequency, and dividing by 60. Though this could be done with multivibrator frequency dividers, I chose to divide by using the 923 flip-flops, which frees me of any misadjustment problems that give an output frequency other than 1 Hz. This frequency divider also provides a 30 Hz and 10 Hz output.

Having these four frequencies, then (60, 30, 10, and 1 Hz), we arrive at the second section, which is the key part. This is the **interval switcher**. It consists of two 923 flip-flops hooked up in a special way. Here is how it works:

Both flip-flops are reset by the **count** button so that there is no voltage output at either pin No. 7. Upon releasing the **count** button, the next 1 Hz wave that goes to zero at the input of flip-flop 1 (FF1) will cause pin No. 7 to go positive; and pin No. 5 will go from its original positive, to zero (these two pins are always at opposite states). When this happens, the count will begin to be registered on the count indicator. At the end of the next cycle applied to FF1, it will switch over to the state that it was at in the beginning; namely, voltage on pin No. 5, but none on pin No. 7. This voltage on pin No. 5 will then cause the count to discontinue, with the last count registered on the count indicator. To prevent the next cycle from counting some more, another flip-flop, FF2, is so connected with FF1 pin No. 3 that the count is locked up until the **count** button is pushed once again, whereupon the whole process starts over.

An additional feature is the switch that can select various submultiples of 60 Hz so that a count may be taken for 1/60, 1/30, 1/10, or 1 sec. This means that the actual count will be the reciprocal of time counted, times the count registered. For example, a count of 1000 on the count indicator taken in 1/30 of a sec would equal an actual count of 30,000.

The third main section is the **gate**, which allows the input to be coupled into the counter for the interval selected. This gate consists of one-half of a Fairchild 914 dual 2-input gate. It consists of two transistors with a common collector resistor and separate base leads. When a positive voltage is applied to one transistor, it draws current; and the resulting current will show almost no voltage at either collector, shutting down the other transistor in the IC. When the count is not to be taken, a positive voltage from FF1 of the interval switcher is applied to one of the bases, thereby nullifying any signal that might be coming to the base of the other transistor in the gate. When a count is to be taken, the interval switcher goes to zero, allowing the count to go to the count indicator.

The fourth and fifth sections are similar, except that the fourth is better suited for the 60 Hz input; and the fifth is a pulse shaper, as is required to operate the flip-flop. They are monostable oscillators that give a pulse of extremely fast rise time. The duration of the pulse is governed by the value of the capacitor. The higher the value of the capacitor, the better the monostable will work at a low frequency, hence, the 0.02 uF in the 60 Hz shaper. On the other hand, a low value must be used for a high-frequency input, hence, the 0.001 in the input shaper so that the unit will go up to 100 kHz. If you wish to measure a higher frequency than this, change this capacitor to a smaller value. To get better low-frequency response, 0.1 uF will provide response from several hertz to 20 kHz.

Construction of Count Controller

In general, the count controller is built and wired similarly to the count indicator. However, the wiring is a bit more involved to achieve the moduli of 3 and 10, so care must be taken and the completed unit carefully checked for errors.

Power Supply

This unit must supply three voltages, none of which is very critical. First of all, there must be 3.6V to operate the various ICs. I have operated this unit from flashlight batteries—the voltage dropped to 2.3V before operation became erratic from one stage. At any rate, try to keep the voltage ±10 percent of 3.6V and you will be assured of reliable operation. This power supply is only my way of arriving at the needed voltages; there is a wide latitude of possibilities. In particular, the 60 Hz supply voltage to the frequency divider flip-flops is very noncritical due to the limiting nature of the pulse shaper it will feed into. Except for the 60 Hz voltage, the whole unit can be operated from dry cells.

USING THE COUNTER

After the normal checks for errors, hook up the power supply and check for proper voltages. If desired, 3V from batteries may be used for the ICs, and another 3V supply may be used for the lamps.

By pushing the **count** button, with the switch on **x1**, you should get the No. 1 lamp to light after a pause of 1-2 sec. with no input. If this happens, things are working. If not, then the 60 Hz shaper is adjusted incorrectly. The easiest and best way to adjust this is to put a scope on pin No. 2 or No. 6 of the 60 Hz shaper and adjust the 10K pot until you have a dependable signal. You will find that this adjustment is not at all critical, and you may hit it accidentally. When it's off, there will be no output appearing at the pins mentioned. Another way, not requiring a scope, is to adjust until you do have the No. 1 lamp come on every time with no input. (This is a feature of this circuit, which tells you that the counter is working, but that

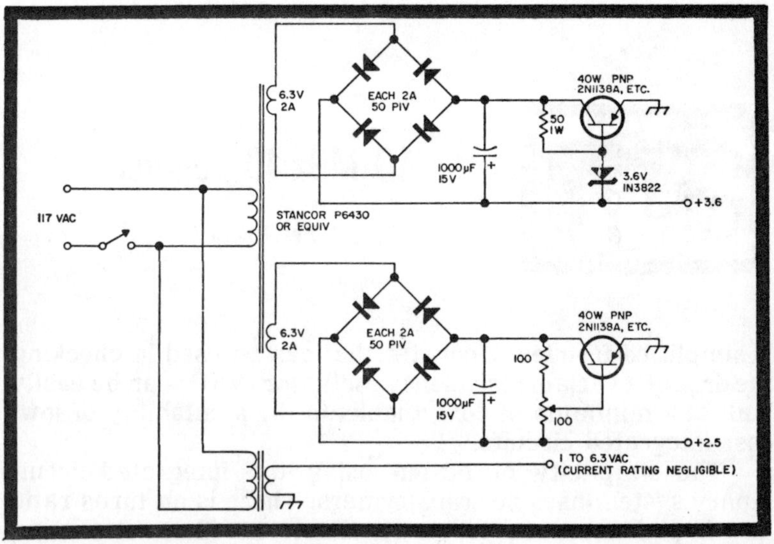

Fig. 2. Power supply for the integrated-circuit frequency counter. Two sources of regulated B+ are provided; one for the ICs, the other for the lamps and lamp drivers.

you do not have sufficient input or that the input is not hooked up properly.)

After you have achieved a reliable **1** count, proceed to couple in a sinusoidal audio signal, preferably around 1kHz at about 1V rms. Push the **count** button, and see if you get a count. If you don't, then adjust the 10K pot in the input shaper section until you do. You will find that it is easier to check the unit if you have a known frequency and watch the counter count it. After getting a correct count, back off on the input ac voltage, and adjust the input shaper's potentiometer until you have the most sensitive setting. With my unit, I get an input sensitivity of 0.1V from 700 Hz to 100 kHz. From 700 Hz to 20 Hz the minimum input requirements rise slowly, requiring 1V at 50 Hz and, finally, 2.5V rms at 20 Hz. It does not overload up to 8V. Where possible, the input should be coupled directly rather than through a decoupling capacitor. When the capacitor is used, the minimum input requirement at low frequencies goes up.

The lights are automatically reset to **0** when the **count** button is pushed. The purpose of the **reset** button is to extinguish the lights for standby.

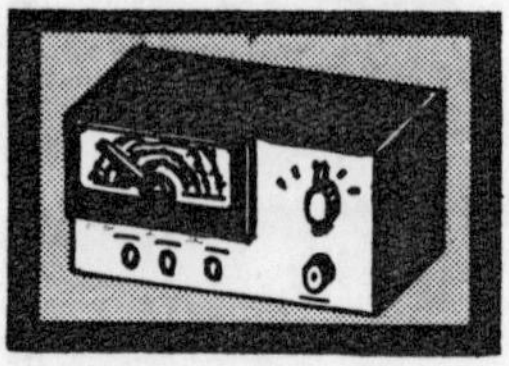

10 MHz IC Counter

Wes Votipka

A simplified frequency counter that can be used in checking the drift of a variable frequency oscillator (VFO) can be easily built at a minimum of cost, thanks to the availability of low-cost integrated circuits.

The simplicity of design using the integrated-circuit binary system uses no transformers. There is no turns ratio and impedance ratio to worry about; no filament to run, shield, and bypass at each end; and no ground loops that always seemed to be present in equipment built in the past.

The basic block diagram for the digital frequency counter is shown in Fig. 1. A standard frequency is needed for a reference, a pulse-shaping network for the input signal, a counting and display system, a power supply, and a control (program generator) to sequence everything properly.

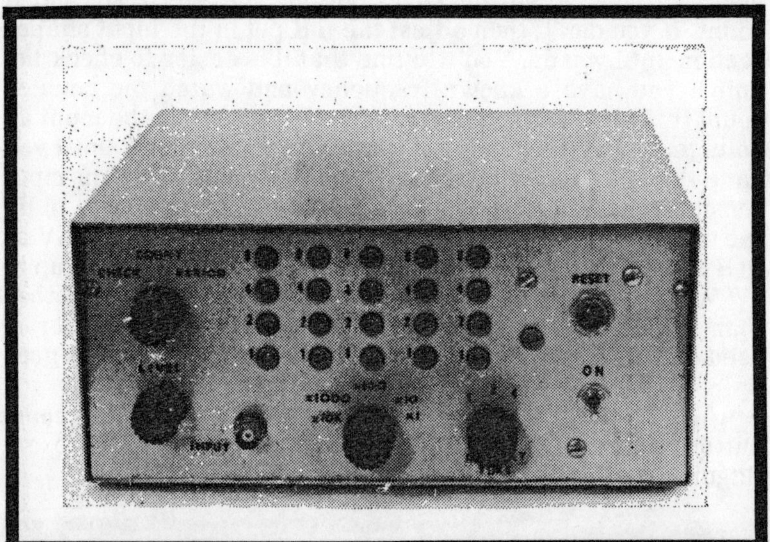

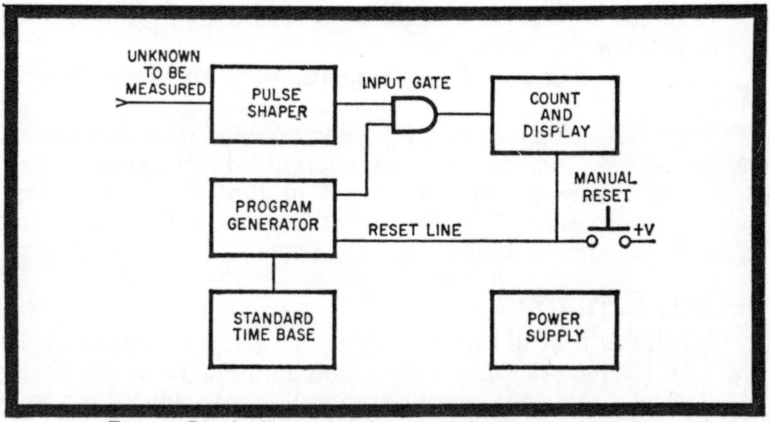

Fig. 1. Block diagram of typical frequency counter.

The final design must meet these requirements:

1. Count frequency to at least 10 MHz.
2. Have reasonable accuracy, 0.001 percent or better.
3. Use low-cost and readily available integrated circuits.
4. Have a line-operated, regulated power supply, with options for use with an auto battery.
5. Have a 5-digit readout with panel lights, because Nixie tubes are too expensive and require additional integrated circuits for decoding for a decimal readout.

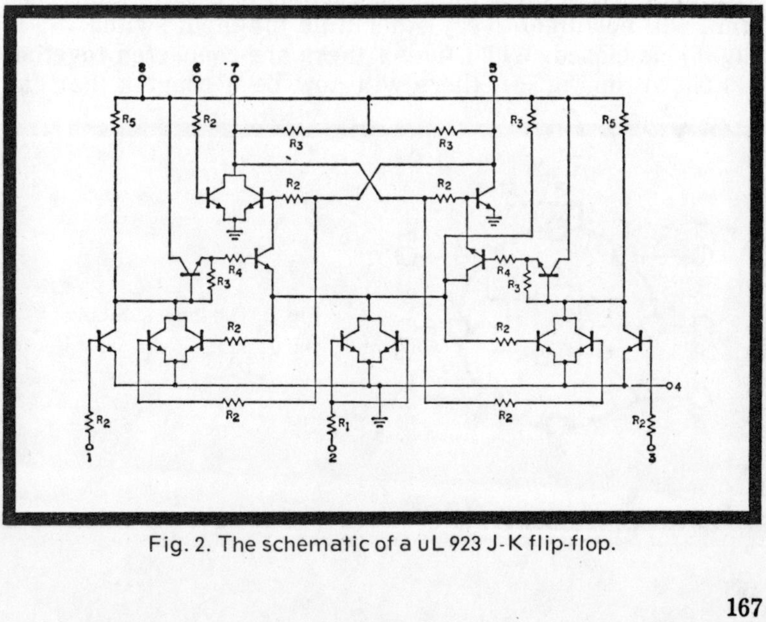

Fig. 2. The schematic of a uL 923 J-K flip-flop.

6. Have a variable readout display time.
7. Count period for additional flexibility.
8. Self-check.
9. Use a PC board for repeatability and trouble-free wiring.

The integrated-circuit frequency counter to be described meets all of the above requirements and exceeds the minimum frequency specifications in that it has a direct readout to 20 MHz. The total cost of new, off-the-shelf integrated circuits is approximately $120.

COUNTER THEORY

First, it's best to examine the basic parts of a counter to see how they work, and then connect them to make them count. The basic unit (counter module) must contain some means of sensing a change at its input. The module should be able to store or hold this information so that it can be displayed on a readout device. The unit chosen for this function is a flip-flop, a device that has only two output states, a 1 and a 0. Figure 2 is the schematic of a uL 923 J-K flip-flop.

Figure 3 is the logic symbol of the uL 923, and the truth table is outlined in Fig. 4. A lamp is used in series with a transistor connected to ground, as a lamp driver, and when connected, produces the counter module shown in Fig. 5.

When connected in this manner, the transistor would draw excessive base current, so a 470-ohm resistor is added in series with the base to limit the base current to a safe value. The lamp will now light every other time the input switch (S1 in Fig. 5), is closed. When five of these are connected together (as shown in Fig. 6), there will now be a counter that can

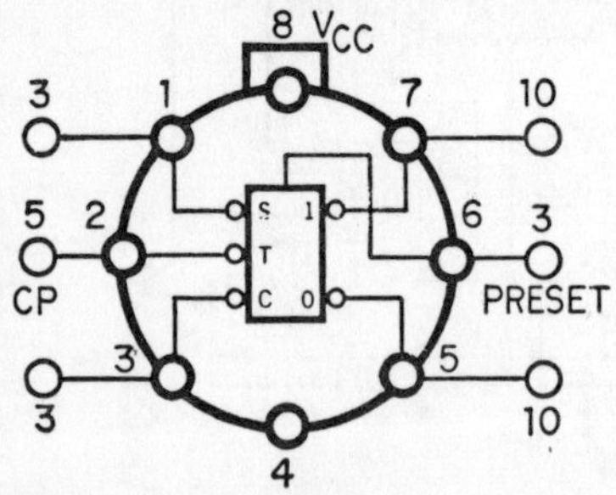

Fig. 3. The logic symbol of the uL 923.

FUNCTIONS

SET (1)	CLEAR (3) t=n	OUTPUT (7) t=n+1
H	H	X^n
H	L	H
L	H	L
L	L	$\overline{X^n}$

H = HIGH X IS THE OUTPUT STATE AT TIME n

L = LOW A HIGH ON PIN 6 WILL PRESET OUTPUT PIN 7 LOW

Fig. 4. Truth table for uL 923.

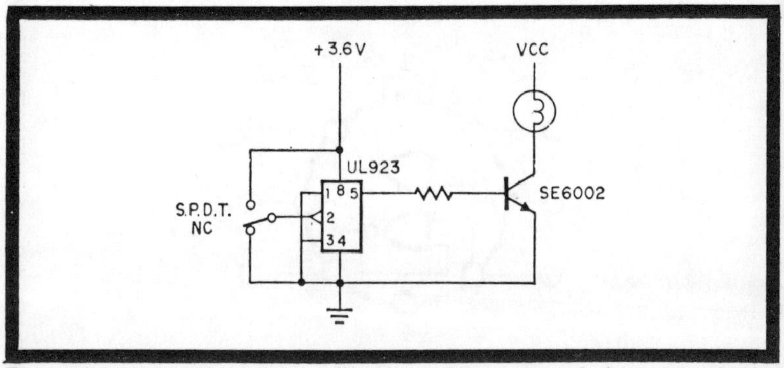

Fig. 5. A lamp used in series with a transistor connected to ground as a lamp driver produces the counter module shown.

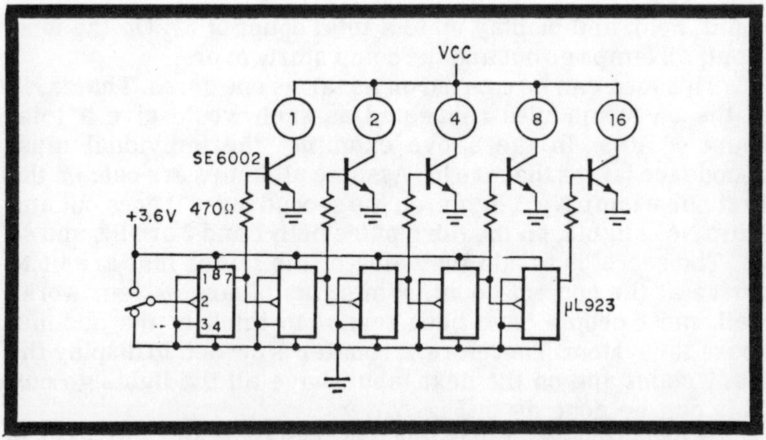

Fig. 6. Five counter modules connected together can count, hold, and display a total count of 31.

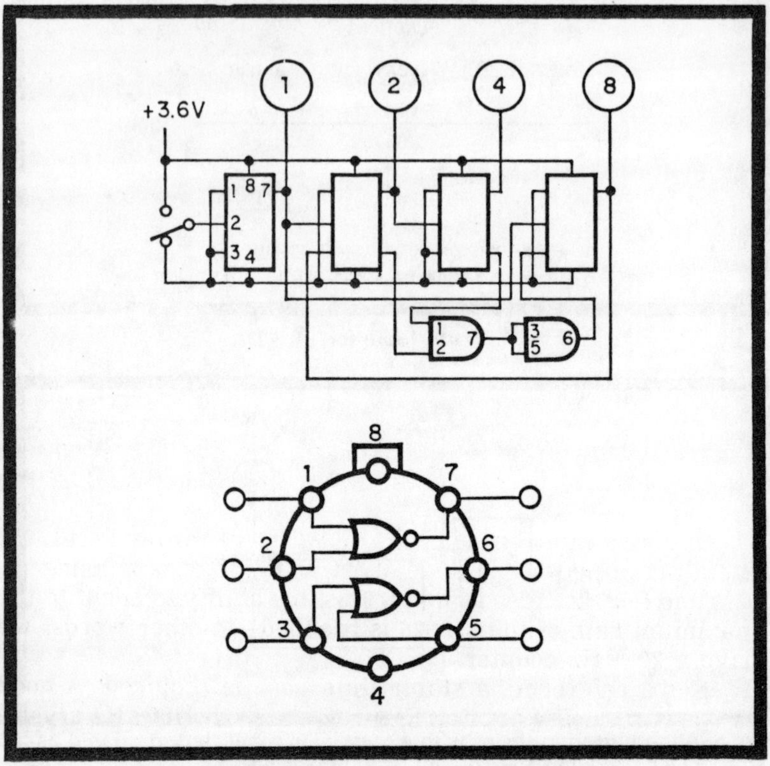

Fig. 7. In A, a decade divider; in B, the uL 914.

count, hold, and display up to a total count of 31. On the next input, all lamps go out and the count starts over.

This idea can be carried on as far as one cares. That is, 12 of the basic modules connected as such would give a total count of 4095. In the above example, the individual must decode the lamps that are lit. Assume all lamps are out: on the first pulse lamp No. 1 lights, on the second pulse it goes out and lamp No. 2 lights, on the third pulse both 1 and 2 are lit, and so on. The secret is to add the value of the lamps that are lit to arrive at the correct count. While this binary system works well, most people have been taught to think in the decimal (base 10) system. Therefore, a counter is needed to display the No. 9 count and on the next input, have all the lights go out. This can be done as in Fig. 7A.

In order to accomplish this, four flip-flops and a uL 914 are used (Fig. 7B) with dual 2-gate inputs connected to form a decade (divide-by-10) counter.

Table 1. Truth Table for 5-Decade Counter.

Table 1

		Lamp Number			
Decimal	Input	#1	#2	#3	#4
0	0	0	0	0	0
1	1	*	0	0	0
2	0	0	*	0	0
3	1	*	*	0	0
4	0	0	0	*	0
5	1	*	0	*	0
6	0	0	*	*	0
7	1	*	*	*	0
8	0	0	0	0	*
9	1	*	0	0	*

The truth table for this (Table 1) is the proof of the operation. By connecting five of these decades together, one has the ability to count to 99,999.

The maximum speed of the uL 923 is 2 MHz, and a uL 926 has a maximum speed of 20 MHz, with one decade using uL 926s and four decades using uL 923s, the ability to count to the maximum rate of the uL 926 is reached—in other words, we have a 20 MHz counter.

For a reference, a stable time base is required. A good choice is a 100 kHz crystal. Figure 8 shows the 100 kHz crystal oscillator used.

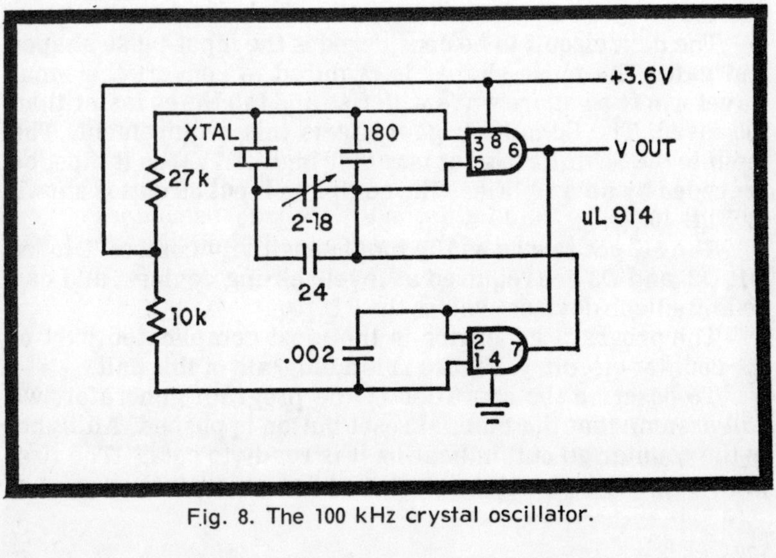

Fig. 8. The 100 kHz crystal oscillator.

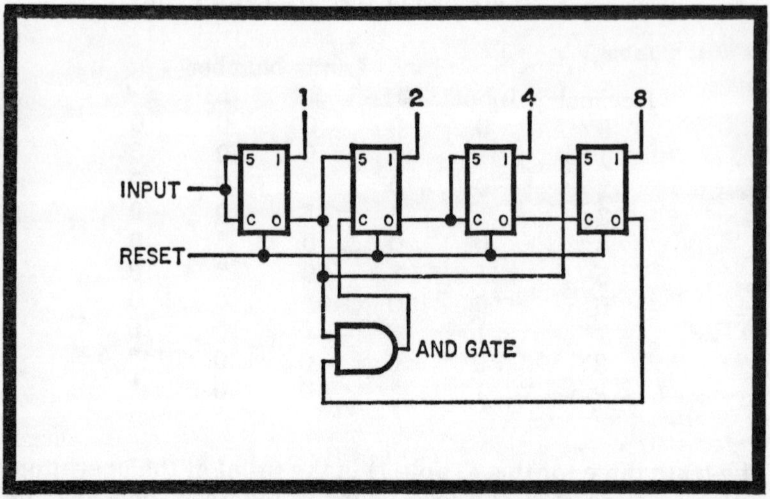

Fig. 9. The logic diagram of the uL 958 decade divider.

The uL 914 is connected as a multivibrator, with the crystal connected in one of the feedback paths. The output of the crystal oscillator has to be reduced to lower frequencies to be useful. The uL 958 decade divider is used here in the interest of space. The uL 923 decade divider of Fig. 7 could be used as well, if space is not a prime consideration. The logic diagram of the uL 958 is shown in Fig. 9.

Five of these decade dividers are required in this system in addition to three uL 923 flip-flops to give the required timing pulses.

The next circuit to be considered is the input-pulse shaper and gate. The pulse shaper is required to convert the input waveform to a square wave with rise and fall times faster than 200 msec. The Schmitt trigger meets this requirement. The input to the Schmitt trigger is rather high (1.7V), so it must be preceded by an amplifier. The complete input circuit is shown in Fig. 10.

The 5K pot serves as the input sensitivity control. Diodes D1, D2, and D3 are required as level-shifting devices, and can be any silicon devices such as the FD 100.

The program generator is the most complicated part of the counter circuitry. Figure 11 is a diagram of this unit.

To describe the operation of the program generator, we will assume that the manual reset button is pushed. All lights in the counter go out, indicating it is ready to count. The first positive-going pulse (from the time base generator) on pin 2 of gate G1 drives its (output) pin 7 in a negative direction, toward ground. This negative pulse causes flip-flop 1 to change state, from a "high" at pin 5 to a "low" at pin 5; the low on pin 5 allows the input gate in the pulse shaper to pass the pulses to be counted to the counter. At the end of the first pulse, G1's pin 2 is no longer held at a high level (approximately +3V), and returns to ground, allowing pin 7 to go high. As the uL 923 flip-flops do not toggle on a positive-going pulse, flip-flop 1 does not change state, and the pulse shaper gate is enabled (held open) to allow pulses to pass. When the second pulse from the time base generator, arrives at the input to G1, it again drives G1's pin 7 low, which causes flip-flop 1 to change state. When this happens, pin 5 of flip-flop 1 is high; therefore, the pulse shaper gate is blocked, and no more pulses are allowed into the counter. At the same time pin 5 goes, pin 7 is driven negative. These actions cause flip-flop 1 to change state with flip-flop 2, pin 7 going from low to high. The high on pin 7 is applied to pin 2 of G1, which inhibits, or blocks, its input; therefore, the count is now displayed until a reset pulse is initiated. Gates G2 and G3 are connected as a one-shot to control the gate lamp and D1, the gate lamp driver. This lamp is lit when the pulse shaper gate is enabled (opened) to let pulses into the counter. The lamp stays lit for approximately 200 msec. The flashing gate lamp is an indication that the counter is either ready or is counting, depending on the presence of a signal at the input to the pulse shaper.

Gate G4 and driver D2 are also connected as a one-shot to automatically reset the entire frequency counter, depending

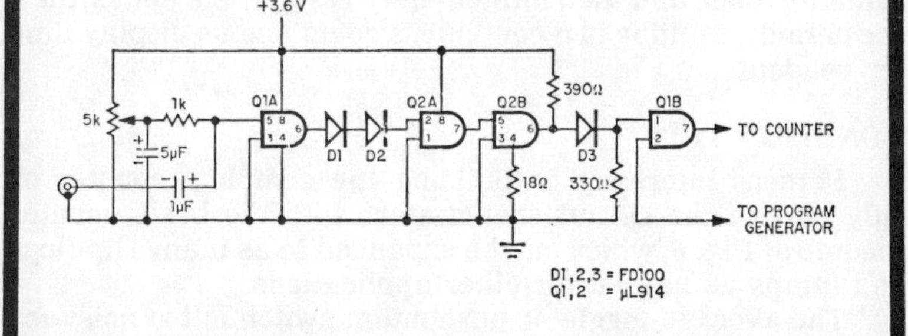

Fig. 10. The complete input circuit to the Scmitt trigger.

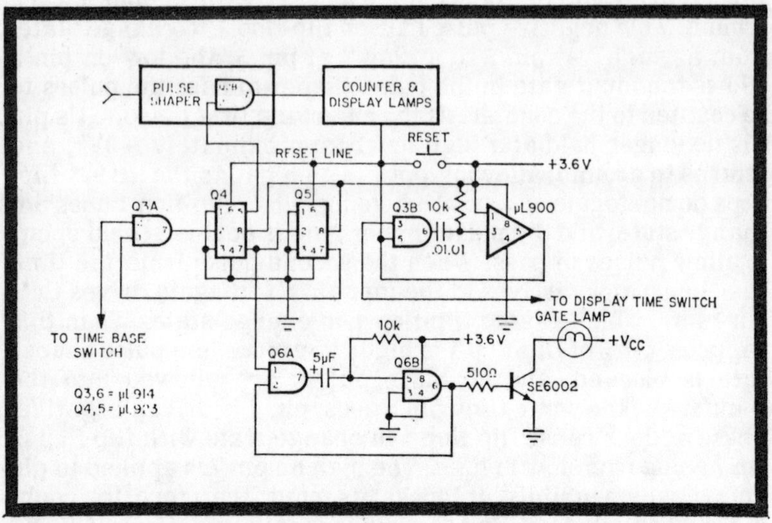

Fig. 11. Diagram of the program generator.

on the position of the **display time** switch. In this design, the display time is 1, 2, and 4 sec. The actual display time is the time selected by the **display time** switch, less a 2 pulse time selected by the time base switch. As an example assume that the **time base** switch is set on 1 kHz (1 msec) and the display time is 1 sec. The count will be for 1 msec and the display for 999 msec. This is very close to the time selected by the **display time** switch, but if the count is for 1 sec, as would be used to count very low audio frequencies, then the **display time** switch should be set for either 2 or 4 sec display for a reading; because, in the 1 sec display time position, the counter will count for 1 sec and then immediately reset at the end of the 1 sec period, resulting in a continuous count and no display time for readout.

CONSTRUCTION

Persons interested in building the complete counter or only part of the circuit should start with the basic counter module of Fig. 6, which can be expanded to as many flip-flops and lamps as needed for other applications.

The average toggle or pushbutton switch is too noisy for this application, so one should not try to use one to control the basic counter module. The one-shot portion of the gate lamp circuitry of Fig. 11 is shown in Fig. 12, and should be used in place of the switch to drive the basic counter module.

The standard coding used on the uL 900, uL 914, and uL 923 is such that pin 4 is ground and pin 8 goes to +3.6V. On the uL 926, pin 5 is ground and pin 10 goes to +3.6V.

Another point to remember is that all unused input pins must be grounded for noise immunity. It should be noted that all pin numbering is done from the top. Remember when working from the bottom of the board that ICs are numbered like the tubes we are familiar with. After becoming familiar with the wiring and operation of Fig. 6, change the layout to the wiring of Fig. 7 and check the proof of the truth table. This table is an informative display, and is the key to the readout of this frequency counter.

Several of the basic counter modules may be joined by connecting pin 1, the carry output of the first module, to pin 2, the count input, of the second unit. These connections should be repeated as additional modules are added to the existing ones. By jumping pin 6 of all of the uL 923s together and adding a spring-loaded normally open switch in series with this line to the +3.6V line, all units may be set to 0 by momentarily pressing the switch to turn off all the lamps.

The next thing to build is the 100 kHz crystal oscillator shown in Fig. 9. The operation of the oscillator depends on the tolerance of the 0.002 uF capacitor, and this value may need to be adjusted slightly to insure crystal control of the oscillator. With the crystal out of the circuit, the output at pin 7 should be a square wave with a free-running frequency of about 150 kHz. This can be checked on a receiver. The next step is to put the crystal back in the circuit and adjust the trimmer to zero beat with WWV when slightly coupled to the antenna. A uL 914

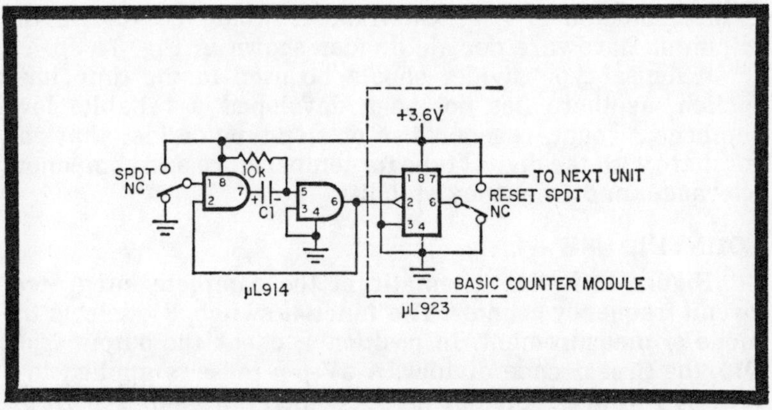

Fig. 12. Noiseless switch used to control the basic counter module.

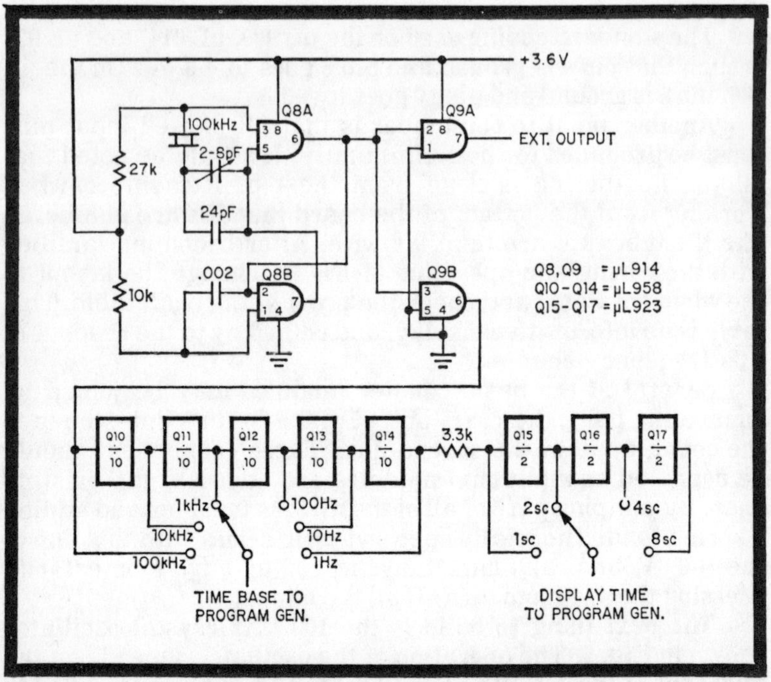

Fig. 13. A uL 914 used as a buffer amplifier to prevent overload of the oscillator.

should be used as a buffer amplifier to prevent loading of the oscillator as shown in Fig. 13.

There are several alternatives to consider in place of the uL 958 decade dividers, if one is not limited by space requirements. One alternative would be to use the basic counter module of Fig. 7A. Another would be to use the minimum-hardware decade divider shown in Fig. 14.

A digital-type divider should be used in the time base section, as there has not been developed a reliable, low-component count, regenerative or step-type divider that can compare with the digital type for temperature and component tolerance, and frequency stability.

COUNTER USE

Figure 15 is the schematic of the complete integrated-circuit frequency counter. The **function** switch, S1, selects the mode of measurement. In position 1, check the output from Q10, the first decade divider. A 3V p-p pulse is applied to a resistive voltage divider and coupled through a 0.01 uF capacitor to the input of the pulse shaper. The level is set at approximately 53 mV to allow a setting of the **sensitivity** control to the most sensitive position for future measurements.

Position 2, **freq**, is used for most frequency measurements. Some adjustment of the **sensitivity** control may be needed to handle very small or very large signals. Some consideration should be given the dc level applied to the input jack. The 1 uF 50V input capacitor should be changed in the event the unit is used exclusively with tube-type circuits. However, an external dc-blocking capacitor may be used, and should **always** be used when in doubt.

Position 3, **period**, is the most versatile one. In this mode, the input to the pulse shaper and the program generator portions are reversed. A pulse applied to the input jack opens gate Q 1B, allowing the pulses (selected by the time base switch, S2) into the count register. These pulses are accumulated in the register until the next pulse arrives at the input jack, which inhibits (turns off) gate Q1B. The accumulated count is now displayed for readout. The reciprocal of the readout is the period of time between the input pulses; i.e., time = 1/readout. For example, a readout of 1000= 1/= 1000=1 msec, of 200 = 1/200 = 5 msec, and 25 = 1/25 = 40 msec.

One use of the **period** position is to measure frequency or occurrences too slow to count reliably. The doppler frequency of a satellite as it passes overhead may be measured in this manner. Other possible uses are as a digital readout automobile tachometer, a timing clock at the drag strip, and a chronograph to measure the speed of a bullet.

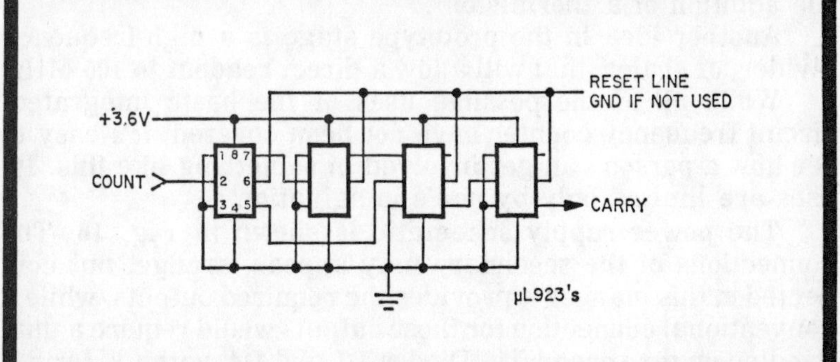

Fig. 14. The block diagram of the standard time base generator.

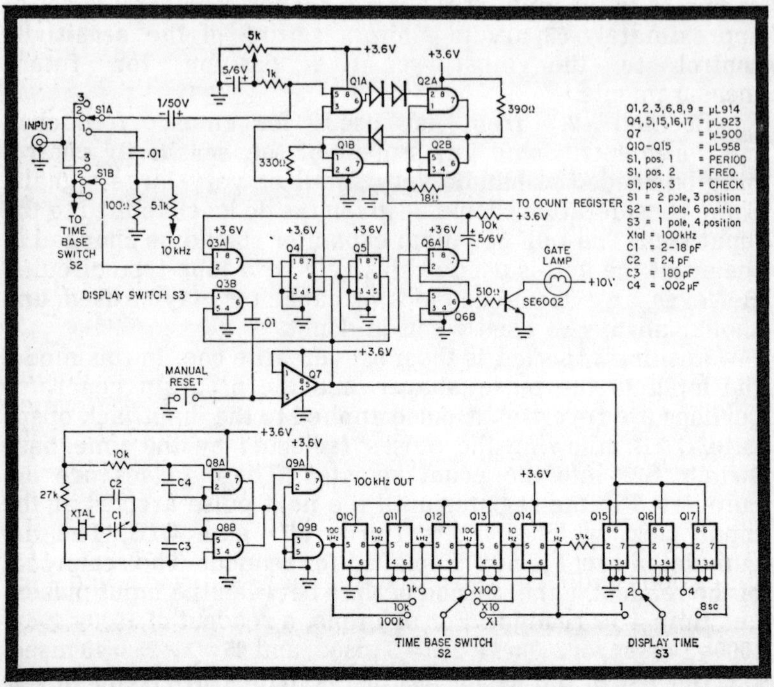

Fig. 15. The schematic of the complete integrated-circuit frequency counter.

A voltage-to-frequency converter now in the prototype stage looks promising. With a conversion factor of 100 Hz/volt, the unit becomes a digital voltmeter with possible uses as a digital VOM and a digital thermometer (by the addition of a thermistor).

Another idea in the prototype stage is a high-frequency divider, or scaler, that will allow a direct readout to 100 MHz.

While all of the possible uses of the basic integrated-circuit frequency counter have not been covered, it's easy to see how a person can get involved in something like this. Its uses are limited only by one's imagination.

The power supply schematic is shown in Fig. 16. The connections of the secondary may appear strange, but connected in this manner it provides the required outputs, while a conventional connection for these outputs would require a dual winding on the secondary. Diodes D1 and D4, with C2, form a capacitor input bridge circuit, with C2 charging to the peak value of the secondary voltage, approximately 6.5V under load. Diode D5 and C1 form a half-wave capacitor input supply with approximately 12.5V output under load. Parts R1, D6, and Q1 form a series regulator for the 10V line to the lamps. Capacitor C3 is needed to suppress the noises generated as the lamps are switched off and on. Parts Q2, Q3, and D7 form a series regulator for the 3.6V line. Resistor R2 is connected between D6 and D7, with D6 as a constant voltage point. Resistor R2 acts as a constant current source for D7. The output of the 3.6V line has excellent regulation. Capacitor C4 is needed for additional noise suppression.

For operation from an auto battery, the power supply is broken at points A and B, and the auto battery adapter is connected to these points.

The photos show the mechanical layout used in the model. The enclosure, the front panel, and the subchassis are formed of copper-clad board. The subchassis is soldered to the front panel after the holes are drilled for the switches and lamps. All components of the power supply are mounted on the aluminum channel, which is bolted to the front panel and the enclosure bottom, resulting in a very light but rigid assembly. The dimensions of the enclosure are 4 in. high, 9 in. wide, and 9 in.

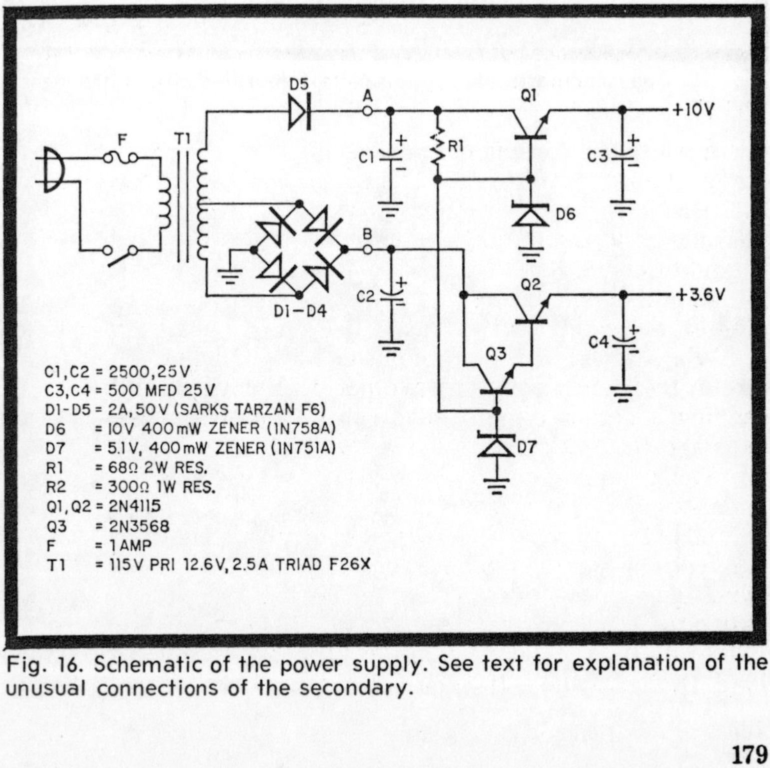

Fig. 16. Schematic of the power supply. See text for explanation of the unusual connections of the secondary.

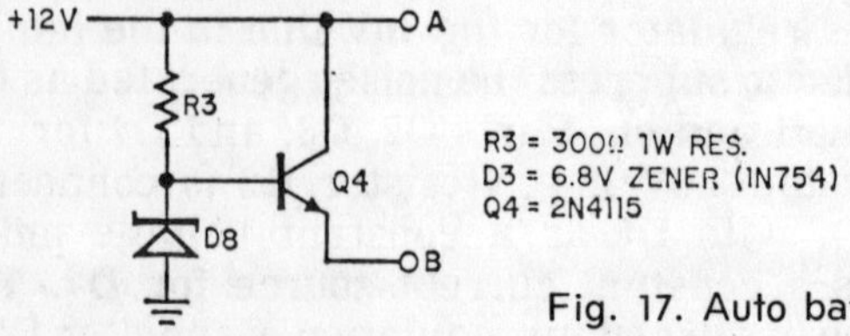

Fig. 17. Auto battery adaptor.

long, with a net weight of 5 lb. While there may be frequency counters on the market with better appearance, better readout systems, higher count ability, and oven-controlled crystals, their cost puts them out of reach of the serious experimenter. This unit shows what can be done, without a large engineering effort, to utilize readily available ICs in a project of this magnitude.

A Modern VHF Counter

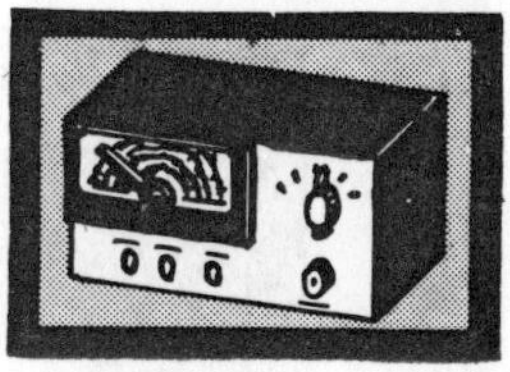

Peter A. Stark

Although a number of frequency counter designs have appeared in magazines and elsewhere, we decided it was time to come up with a new design. We have gone out of our way to make the building of this counter as easy as possible, and yet to produce a counter of really top-notch performance. First, we will give a complete description of what the counter is and what it can be used for, as well as the complete parts list so you can start collecting parts if you decide to go ahead. Next, we will give the complete logic diagrams and theory of operation. Then we will have the printed-circuit board layout, parts layout drawings, and construction and operating information.

COUNTER FEATURES

And now a little information about the counter:

1. The basic counter measures up to 20 MHz, although many ICs go past that. One of our prototypes goes up to 35 MHz, and the other goes to about 40 MHz. But that's only the basic counter—an optional VHF prescaler mounts on the same board and extends coverage to above 200 MHz. Depending on the ICs, you may get operation higher than that. We've had our prototype counting at the upper limit of our signal generator, which is 250 MHz. With a simple modification, the scaler range can be increased to typically 320 MHz, though at reduced sensitivity.

2. It is easy to build. Except for the power supply, all other small parts fit onto one 7 x 10 in. printed-circuit board. This means there's a minimum of external wiring and less chance for error. The PC board layout, etched boards, and complete parts kits are available to make the job even easier.

3. The resolution is within 1 Hz with the basic counter up past 20 MHz, and within 10 Hz with the VHF scaler. The accuracy depends on the quality and type of reference crystal

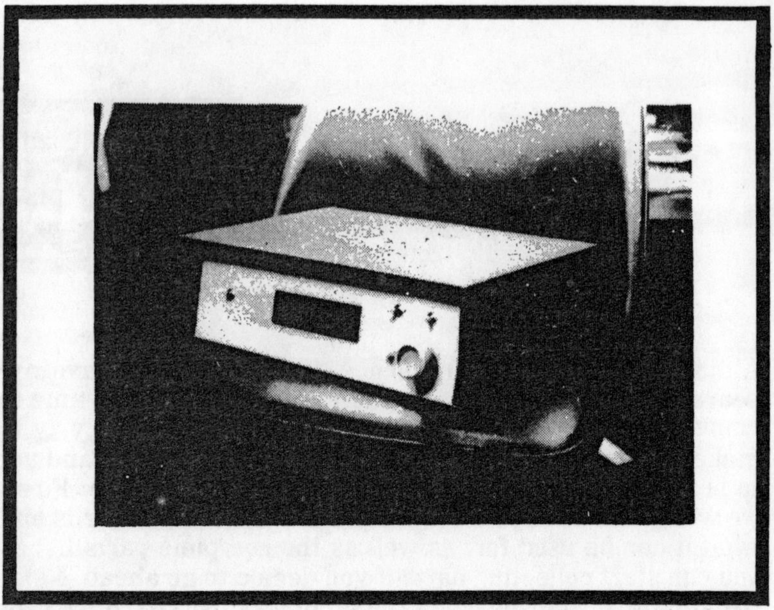

used; the circuit allows you to zero the crystal against WWV. With the AT cut series-resonant crystal suggested, the specs say that the drift should be less than 0.5 parts per million (ppm) per month, and the temperature coefficient should be about 0.077 ppm per degree Celsius. In our counter, the crystal is mounted close to the edge of the PC board with the idea that someday we will place it in a crystal oven. Let's face it, the measurement accuracy is only as good as the reference crystal accuracy, and all the really expensive commercial counters have crystal ovens. Although we will not describe an oven in this article, we suggest you keep it in mind.

4. The design is really up to date. It uses the latest TTL and ECL integrated circuits. To simplify construction, complex MSI (medium-scale integration) ICs are used to reduce the number of interconnections. Although the board is laid out for Numitron readout tubes, light-emitting diode (LED) readouts can be used to provide state-of-the-art reliability as well as really snazzy looks.

5. The price is really reasonable. Complete with all parts, including cabinet, line cord, and all the other necessary things so often forgotten in parts estimates, the entire counter with the VHF prescaler will cost you about half of what you'd pay for the cheapest commercial units, including kits.

Here's how the price for the various portions of the counter breaks down.

1. Basic counter circuitry including all ICs and small parts, but not including VHF prescaler, readout indicators, and power supply—about $65.

2. Power supply, about $15.

3. VHF prescaler, (optional) about $25.

4. Readout indicators and associated circuitry. The price depends on which indicator type you choose. Numitron readouts will add about $25, miniature incandescent readouts (sometimes called Minitrons) will cost about $20, and LEDs will cost about $35. (Prices from one source dropped to $6 per LED at the time of writing, and we expect other dealers to follow suit.) Though the Minitrons are the cheapest, we don't really recommend them, since the $5 saving over Numitrons doesn't justify the added work. Numitrons are viewed from the side, and plug right into the PC board. In this way you can read the digits from the front of the board. The Minitrons, on the other hand, are viewed from the top. This means you need a separate PC board, mounted at right angles to the main board and in front of it, to provide a convenient readout. Those two boards will connect to each other with 38 wires! Incidentally, the LEDs mount the same way, but that's the price of progress.

5. Cabinet, input connectors, test leads, assorted hardware, etc.—about $15.

These prices assume you have to go out and buy all new parts, though not necessarily from a dealer who charges list price. Quite a few of the ICs are available at below cost from a number of mail order outlets around the country, and we used these prices in calculating the total price.

If you have a modest-sized junkbox, you may be able to shave these prices a bit, but don't expect miracles. Unless your junkbox is very modern, it probably won't contain too many useful parts aside from things like switches, connectors, fuse and line cord, and the like. Still, it seems quite reasonable that with a little effort you could build the complete counter with Numitron readout but less VHF scaler for less than $100, and for about $125 with it. This is roughly one-half or less of the price of the Heathkit counter or counter-scaler combination. And it isn't much more than some commercial VHF scalers alone.

CIRCUIT OPERATION

Figure 1 shows a fairly complete block diagram of the entire counter. The individual diagrams for each block will be shown later, and the block diagram gives the figure number for the detailed diagram.

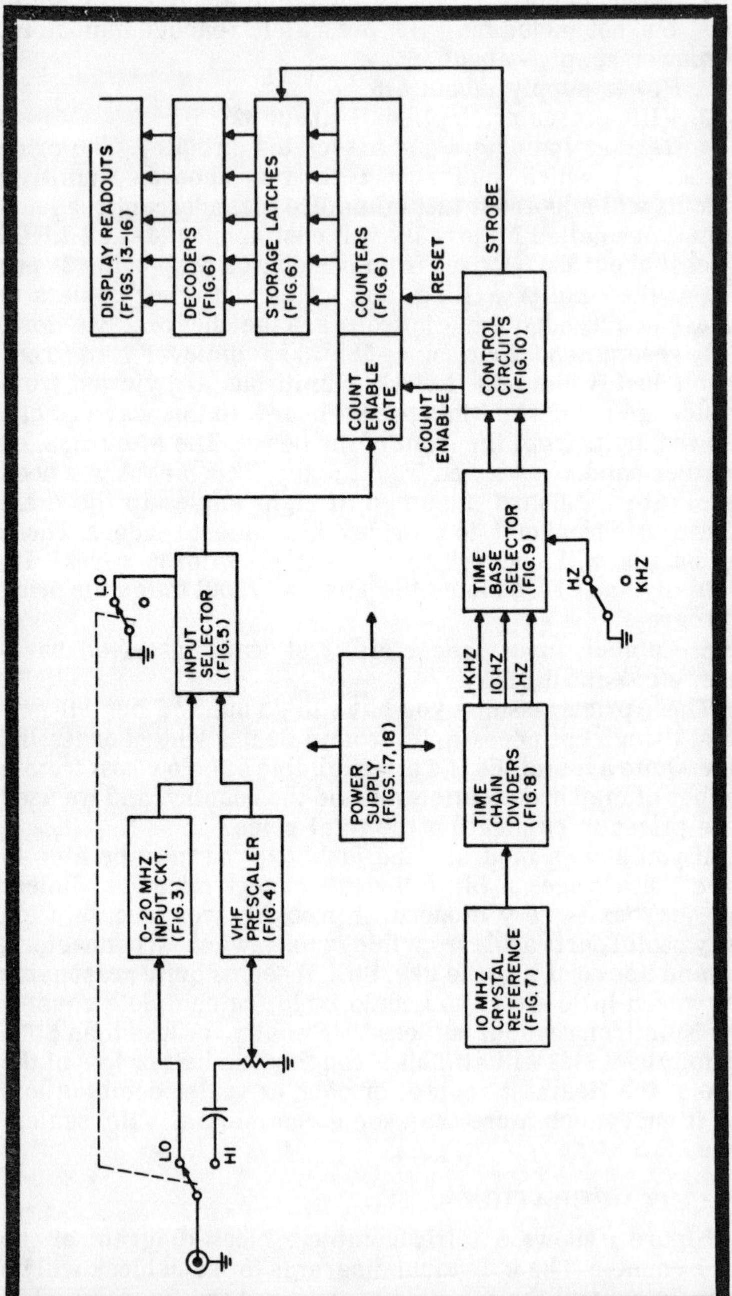

Fig. 1. Counter block diagram.

The input signal is fed by the "hi-lo" switch to either a 0-20 MHz input circuit, or to the VHF prescaler. The low-frequency input circuit just converts the input signals into digital pulses, while the VHF prescaler divides the input frequency by 10 to bring it into the range of the rest of the counter circuitry. The outputs of these two circuits are sent to the input selector, which selects one input and sends it on to the **count enable** gate, which is part of the counters. The count enable gate is opened by the control circuits for a specified length of time, and the number of cycles during that time is counted by the counters. The count is then transferred into the storage latches, which hold the count while the counters take the next count, to produce a nonflickering display. The output from the storage latches goes to the decoders, which then feed the display readouts. The display shows a 5-digit number, and has an overrange light if the count exceeds five digits.

The time reference for the count enable gate comes from a 10 MHz crystal oscillator, whose output is divided by seven decade counters in the time chain divider circuit, to produce 1 kHz, 10 kHz, and 1 Hz pulses, which go to the time base selector. The selector chooses the correct frequency pulse train and sends it to the control circuits.

The control circuits govern the basic timing of all functions of the counter, and generate the count enable, strobe, and reset signals, in that order, over and over. Thus, the counters are allowed to count for a certain amount of time, then the count is transferred (strobed) into the storage latches, the counters are reset back to zero, and then the count enable comes again for the next count.

Measurement Periods

One of two measurement periods is selected by the **Hz-kHz** switch. Assuming that you are not using the VHF scaler, the **Hz** position of the switch results in a count enable signal lasting exactly 1 sec. Thus, the counters count the input frequency for exactly 1 sec, and the 5-digit display then shows the rightmost five digits of the total count, up to a maximum of 99999. If the actual input frequency is higher, the readouts still show the rightmost five digits, but an overrange light indicates that there are more digits to the left. For example, suppose the input frequency is 1,234,567 Hz. Then, the readout shows the digits 34567 and the overrange light is on.

If we want to see the digits to the left, we just move the switch to the **kHz** position, which selects a **count enable** period

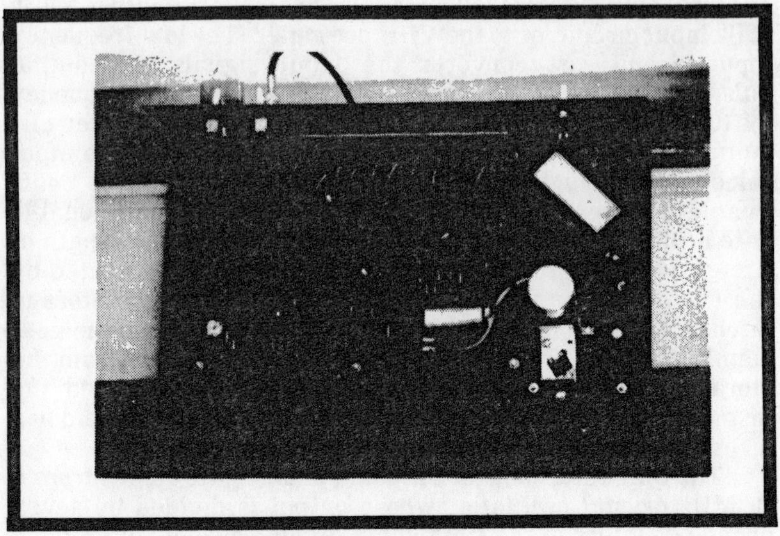

Top view of counter.

of 1/1000 sec (1 msec). If the input frequency is 1,234,567 Hz, then there will be 1,234 cycles in that millisecond, and the readout will show the digits 01234. Put the two readings together, and you have a readout down to the last cycle.

VHF Scaler

Using the VHF scaler, all input frequencies are divided by 10. For example, if we want to measure a 146.940123 MHz signal, the scaler converts the frequency to 14.694012 MHz. In the **kHz** position of the **Hz-kHz** switch we will have a readout of 14694, while in the **Hz** position we will get 94012 with the overrange lamp lit. Thus, we have resolution down to 10 Hz.

While it may seem a bit inconvenient to have to use two readings to get an accurate frequency indication, there are two good reasons for doing it this way. First, to get the frequency in just one reading, we would need an 8-digit readout. That means three more counters, three latches, three decoders, and three indicators. That's an increase of 60 percent in price in that part of the counter. If price is no object, it's easy to add the extra parts yourself.

The second reason is that to get an accurate readout down to the last hertz (or last ten hertz with the VHF scaler) takes a whole second of counting. If you don't need that accuracy, that's a relatively long time. In the **kHz** position we get a new reading every 1/10 sec, which really speeds up the process, especially if the frequency is changing and you want to follow the change.

APPLICATIONS

Before we get on to the parts list, it might be a good idea to look at some applications of the counter. Figure 2 shows some typical ways to connect the counter to a circuit to measure the frequency of a signal.

Figure 1A shows the easiest way—connecting directly into a circuit. The low-frequency input circuit on the counter has an input impedance of about 1M shunted by less than 20 pF, approximately. The VHF scaler has an input impedance of 50 ohms. But avoid overdriving the inputs, to avoid damage. The low-frequency input can stand voltages up to about 100V peak, though it may not count well at inputs over 20V. The VHF input uses a 50-ohm attenuator, and you should make sure not to overdrive the input stage, since this may damage the input

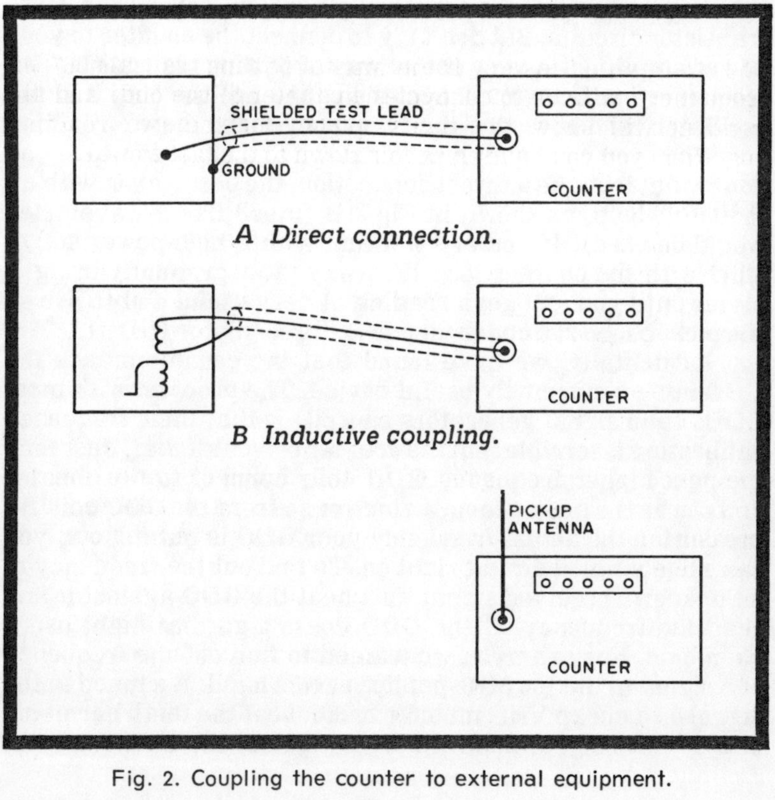

Fig. 2. Coupling the counter to external equipment.

circuit or slow it down. (For highest speed it's necessary to keep the input transistors from saturating, which means keeping the level just above the point where the counter just starts counting.) The best habit here is to start with the attenuator at minimum and turn it up just enough to get a good reading.

It's hard to give a sensitivity figure for both inputs, since we don't have voltmeters capable of measuring at rf frequencies. But it appears that the low-frequency input needs about 200 mV rms, and the VHF input needs about 400 mV. Incidentally, both inputs have input isolation capacitors, so the counter can be connected to a point where a small dc voltage is present. But watch out—don't exceed the breakdown voltage of the capacitor you use! This limits the low-frequency input to frequencies above about 20 Hz; if you want to go below that you will just have to increase the value of the isolation capacitor, or remove it altogether.

Using the direct connection, we've been able to measure the frequencies in tube-type equipment as well as low-level transistor circuits. But don't try to connect the counter to your power amplifier! A very useful way of putting transmitters on frequency has been to connect the counter to the output of the oscillator, and measure the frequency right there.

When you have a high-power stage or a stage in which you don't want to make a direct connection, the best way is with an inductive loop, as shown in Fig. 1B (much like a wavemeter would be used). If you are working with a high-power stage, start with the counter loop far away, and **gradually** bring it closer until you just get a reading. A very useful coil to use as the pickup loop is a coil from a grid dip oscillator (GDO).

Incidentally, we have found that the counter makes the GDO into an extremely useful device. The problem with most GDOs (and signal generators as well) is that their frequency calibration is terrible, but the counter solves all that. Just take the next higher frequency GDO coil; connect to the counter and use as the pickup loop, an inch or so from the GDO coil. By measuring the actual frequency your GDO is putting out, you can align a tuned circuit right on. To find out the frequency of an unknown received signal, just beat the GDO against it and read the frequency. If the GDO doesn't go that high, use a harmonic. For example, we wanted to find out the frequency of a signal in the 160 MHz public service band. We tuned in the signal on a cheap VHF monitor radio, beat the tenth harmonic of a 16 MHz signal from the GDO against the unknown, and read the 16 MHz frequency on the counter. Multiply by 10, and you have the right value—you don't even need the VHF scaler.

Another application along the same line comes when you have a real low-level stage, in which there just isn't enough signal to count the frequency directly with the counter, or in which there are several frequencies at the same time, as in a reflex amplifier. We solved that by plugging an earphone into the GDO, zero beating it against the unknown, and reading the GDO frequency with the counter.

Finally, Fig. 1C shows how to measure the frequency of a transmitter—just connect the counter to another antenna near the transmitting antenna. We've used this approach at 144 MHz, with a 19 in. piece of wire connected to the counter. We've been able to measure the frequency of a 1W handie-talkie about 5 ft away, which shows that the sensitivity is quite good.

PARTS REQUIREMENTS

To end the overall description of the counter, the parts list provides a complete list of everything you need to build the counter. The parts list is divided into five parts. (A through E.)

Part A has all the basic counter parts except for the readouts, the VHF scaler, and the power supply. All of these parts fit on the main 7 x 10 in. printed-circuit board, except for the Hz-kHz switch, which mounts on the front panel.

Part B is the power supply, which provides a regulated +5V ± 5 percent at about 1.5A, and +25V unregulated at about 10 mA. There's quite a bit of room for experimentation here, but make sure that the +5V supply voltage is well regulated. The diodes and filter capacitor for the +25V supply mount on the main PC board, but the +5V supply mounts separately. We did not design our own PC board for the +5V supply, since a number of commercial supplies and kits are available.

Part C describes the optional VHF scaler. Except for the 50-ohm input pot and the "hi-lo" switch, all of these parts also fit on the main PC board, and can be added at any time.

Part D lists the parts for the readouts, with a choice of either Numitron, Minitron, or LED readout. (You could substitute Nixie tubes, but then you'd have to redesign the PC board, since different IC decoderdrivers are needed.) The Numitron readouts mount right on the main board, using 9-pin miniature tube sockets.

Minitron and LED readouts need a separate board to hold the readouts. For the Minitron, you can use perfboard with

Numitron readouts.

holes on 0.1 in. centers. The LED readout, however, needs 150-ohm current-limiting resistors, so we have designed a 3 x 7 in. PC board to hold the five LED readouts, the current limiting resistors, and a single red LED for the overrange indicator.

If you use the Numitrons or Minitrons, you will need a separate 3V lamp power supply, as well as a driver transistor for the overrange lamp. Except for the lamp supply power transformer, components mount on the main PC board.

Finally, part E of the parts list gives some of the miscellaneous parts you will need to complete the counter. As far as the cabinet is concerned, you're on your own. Our prototype (shown in the photos) used a Bud Tilt-a-View

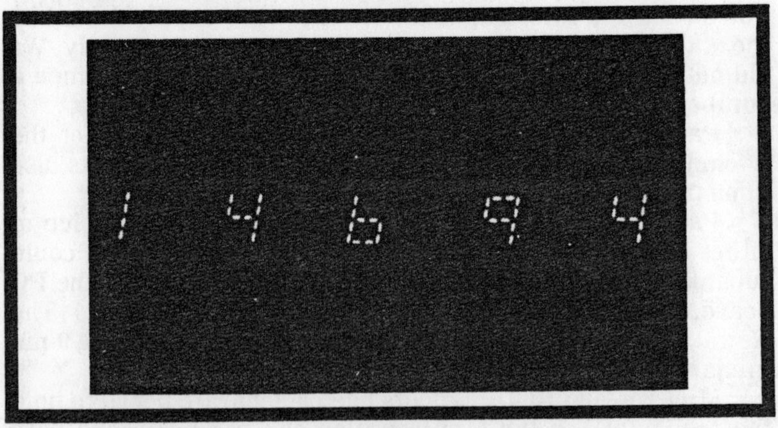

LED readouts.

cabinet (TV-2155), but this cabinet is a lot bigger than you need. We wanted to leave everything open so it would be visible in the photos, but the counter can be crammed into a small cabinet, with the power supply stacked above or below the main board.

CIRCUIT DESCRIPTIONS

Now, we will provide the complete logic and schematic diagrams, and describe how the counter works.

Figures 3 through 18 show the diagrams of the various parts of the counter. To see how these parts fit together, refer back to Fig. 2, the block diagram.

0-20 MHz Input Circuit

This circuit (Fig. 3) is one of the most tricky circuits in the counter, since it must convert a variety of input signals—large and small, simple and complex—into digital pulse signals of just the right voltage and speed to operate the counter. This job is considerably simplified by a Schmitt trigger IC, an SN7413N, designed just to interface TTL digital logic to the outside world.

The low-frequency input circuit starts off with Q1, a 40673 RCA dual-gate-protected MOSFET transistor, which provides high input impedance and a useful amount of gain at a very low price. The input signal from the "hi-lo" input switch is applied to gate G1 through a 100K current-limiting resistor. Because of the diode-protected gates, the 100K resistor allows inputs up to 50V without damage to the FET.

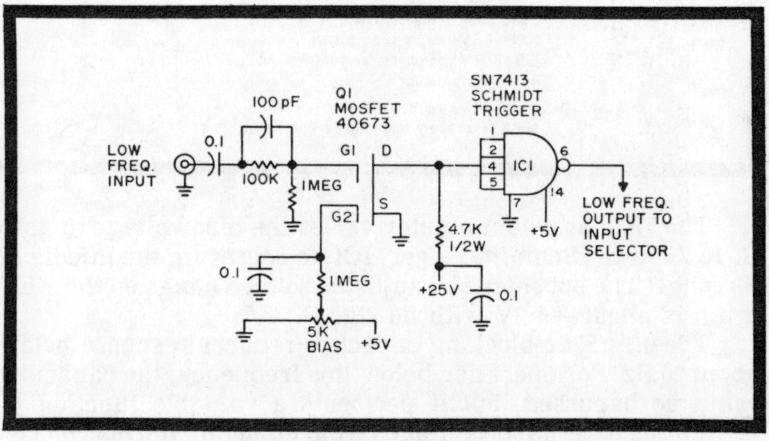

Fig. 3. The 0-20 MHz (low-frequency) input circuit.

FREQUENCY COUNTER PARTS LIST

A. Basic counter – less readouts, scaler, power supply

12	SN7490N	TTL decade counters
4	SN7400N	Quad 2-input TTL nand gate
5	SN7475N	Quad TTL latch
5	SN7447N	BCD-to-decimal TTL decoder/driver
2	SN7473N	Dual TTL J-K flip-flop
1	SN7476N	Dual TTL J-K Flip-flop
1	SN7413N	Schmidt trigger TTL
1		10-MHz AT-cut series resonant HC-18/U crystal
1	1.7–14.1	pf variable air capacitor (Johnson 189-505-5
1	40673	RCA dual-gate-protected FET transistor
2	2N5172	NPN transistor or equivalent
1	7 x 10 in.	printed circuit board – see part 3 of article
1	5K	pc mount potentiometer (CTS X-201-R502B or equiv)
2	220Ω	¼ watt 10% resistor
1	560Ω	"
1	1K	"
1	1.8K	"
1	2.2K	"
2	4.7K	"
1	10K	"
1	27K	"
1	100K	"
2	1 Meg	"
1	4.7K	½ watt 10% resistor
1	33 pF	NPO disk capacitor
2	100 pF	disk capacitors
14	0.01 μF	"
6	0.1 μF	"
450		MOLEX IC socket pins (used instead of IC sockets)
1		SPST switch (Hz-kHz switch)

B. Power Supply

1		Transformer, 6.3V 2 amp and 38V CT 50 mA
2		500 mA 100 PIV rectifier diodes
1		1000 μF 50V electrolytic capacitor
1		SPST on-off switch
1		line cord
1		½ amp fuse and fuse holder
4		2 amp 50 PIV rectifier diodes
1		5000 μF 15V electrolytic capacitor
1		100 μF 15V electrolytic capacitor
1		0.1 μF disk capacitor
1	LM-309K	IC 5V voltage regulator, TO-3 can
1		heat sink for above, minimum 2 x 4 in., finned

The 5K **bias** potentiometer varies the bias voltage to gate G2 to allow the Schmitt trigger, IC1, to operate in the middle of its range. The pot should be adjusted so the voltage on the FET drain is about +1.3V with no signal.

The 0.1 uF dc-blocking capacitor reduces response below about 20 Hz. For operation below this frequency, the capacitor could be bypassed, but it performs a valuable function of preventing external dc voltages from changing the bias on Q1.

		C. VHF Scaler (optional)
1	μ6B958259X	Fairchild ECL amplifier IC
1	μ6B95H9059X	Fairchild ECL scaler IC
1	SN7400N	TTL Quad 2-input gate
32		MOLEX pins
1	2N5771	Fairchild PNP switching transistor
7	0.01 μF	disk capacitor
2	0.1 μF	"
1	50Ω	panel-mount carbon pot (Allen-Bradley JA1N056S500MA or equivalent.
1	2K	p.c. mount potentiometer (CTS X-201-R252B or equiv.)
1	120Ω	¼ watt 10% resistor
2	10K	"
2	330	"
6	220	"
1	4.7K	"
1		DPDT switch (HI-LO switch)
		D. Readout Indicators – choice of D-1, D-2, or D-3
D-1	**Numitron Readouts**	
5	DR-2010	RCA Numitrons
5.		9-pin p.c. mount tube sockets
1*	5V 30 mA	bulb (Sylvania 5ESB or equiv) with matching lampholder and lens
1*	6.3V 1 amp	filament transformer (center-tapped)
2*	1 amp 50 PIV	rectifier diodes
1*	10K	¼W 10% resistor
1*	2N5172	NPN transistor or equivalent
D-2 – Minitron indicators		
5		Minitron readouts
5		16-pin IC sockets
Plus all starred components from D-1 above, and a circuit board.		
D-3 – Light-emitting-diode readouts		
5	LED	Readouts. (Monsanto MAN-1 or MAN-3, Litronix 10A, Environmental LED-700, or equivalent)
35	150Ω	¼ watt resistor 10%
1		red LED overrange indicator
1	220Ω	¼W 10% resistor
50		MOLEX pins
1	3 x 7 in.	p.c. board for LED readouts
		E. Miscellaneous
Cabinet – Input connector - Test leads – Line cord strain relief – Bezel or polarized glass in front of readouts – Hardware.		

VHF Prescaler

The VHF prescaler (Fig. 4) accepts input signals from the low rf range up through about 200 MHz—depending on the ICs you get—and divides the input frequency by 10. This is done by two special-purpose—and expensive—ICs.

Input amplifier IC2 is a high-frequency amplifier that allows the use of quite low voltages. It is not really needed. In fact, to get the scaler to work at 300 MHz and slightly above, it

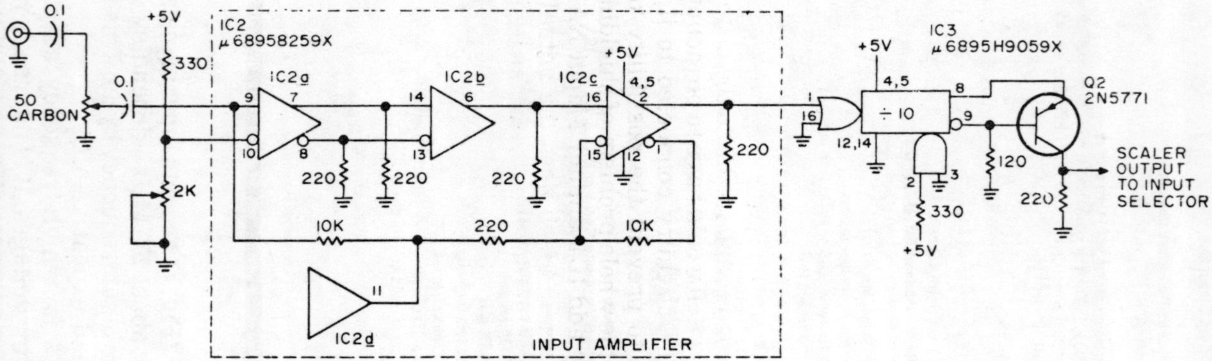

Fig. 4. VHF prescaler. NOTE: to extend scaler range to 300 MHz (with some loss of sensitivity) eliminate all parts in dashed box, and jumper pins 2, 9, and 10 at IC2 socket.

shouldn't even be used, since it limits the high-frequency response. But it is very useful at lower frequencies (below about 180 MHz), and it provides a good buffer for the really expensive IC3, preventing burnout due to high input voltages.

Part IC3 is a special VHF prescaler IC, made by Fairchild just for this use. It replaces four separate ICs used in some scalers, and even at $16 is cheaper than the ICs it replaces. This particular IC is rated to work up to about 320 MHz, and some samples may work even higher than that.

These two ICs are ECL (emitter-coupled logic) ICs and need special interfacing with the rest of the counter, which uses TTL ICs because of their much lower cost. Interfacing is provided by Q2, a level shifter and amplifier.

Input Selector

At first glance, the input selector (Fig. 5) may seem a little overdesigned. It could be replaced by a simple SPDT switch to select the output of either the low-frequency input circuit or the VHF prescaler and feed it to the counters. But to avoid problems, that switch would have to be mounted right on the PC board, near the circuitry connected to it, and away from the input leads to prevent the possibility of feedback oscillations. To avoid the whole problem of shielding the leads and switch, we simply add IC4, a quad 2-input NAND gate IC costing less than 40 cents, which does the actual switching. Now a simple SPST contact on the "hi-lo" switch controls a small dc voltage, which energizes **either** gate IC4b or IC4d. A commercial counter would probably solve this problem by using a 2-deck rotary switch with 5 in. of shaft separating the two decks, with the front deck switching the inputs between the two input circuits, and with the rear deck—neatly

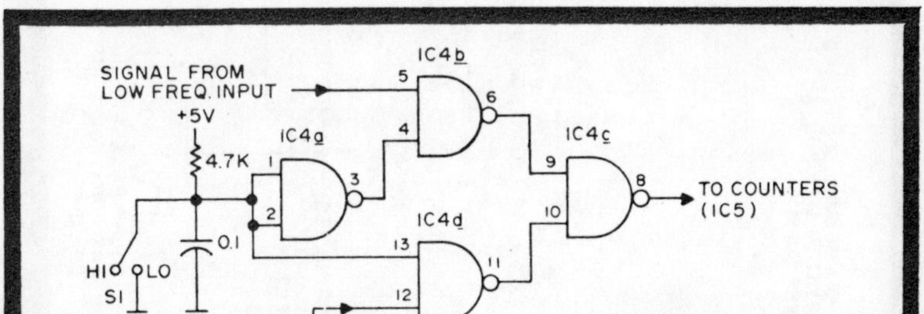

Fig. 5. Input selector.

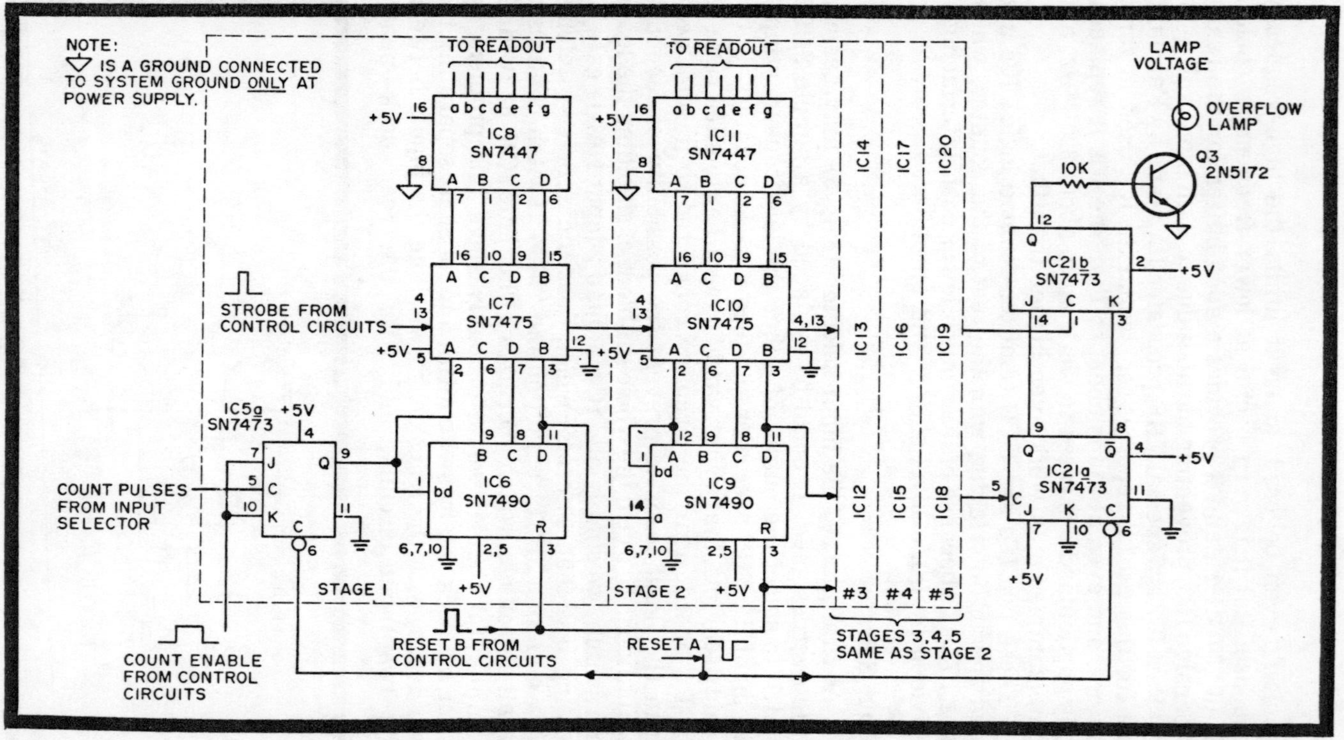

Fig. 6. Counters, latches, decoders, and overflow.

positioned just above the circuit board—controlling the outputs. But for our purposes, using IC4 is a lot more practical, unless you have a switch factory next door.

Incidentally, we are going to use this trick one more time, later, in the time base selector. In this, an SPST switch and two 40 cent ICs do the work of a DPDT switch and five coax leads.

Counters, Latches, Decoders and Overflow

Now we come to the heart of the counter, where the actual counting is done. This circuit is divided into five almost identical stages, one for each digit displayed, plus three extra flip-flops. Most of the work is done by three types of MSI ICs. The five SN7490 ICs (IC6, IC9, IC12, IC15, and IC18) are decimal counters, which count, digit by digit, the actual number of input cycles.

As soon as a count is completed, it is transferred into the five SN7475 latches (IC7, IC10, IC13, IC16, and IC19) when a strobe pulse arrives from the control circuits. These latches act as temporary memories to allow a nonflickering display of the count—even while the counters are reset back to zero by a pulse on the **reset B** line—and then take the next count.

The BCD (binary-coded decimal) digit from each latch is connected to the SN7447 decoders (IC8, IC11, IC14, IC17, and IC20), which translate the BCD code into the 7-segment code needed by the readouts.

The five stages of the circuit are identical, except for stage 1. Part of IC6 (using pins 12 and 14) is instead replaced by an external flip-flop, IC5a. This is done because we need a foolproof way of enabling and disabling the counter for a 1 sec (or 1 msec) interval without disturbing the count or adding extra pulses. This is done by the **count enable** signal applied to the J and K inputs of IC5.

Flip-flop IC21 keeps track of any overflow from the last decade counter, IC18. If the number of cycles counted during the time period exceeds 99,999, IC18 sends a pulse to pin 5 of IC21a, which turns on. Flip-flop IC21b acts as a 1-bit latch, which then remembers this condition, and lights the overflow light through Q3.

Flip-flops IC5 and IC21 are reset by the **reset A** pulse when the counters start on a new count.

10 MHz Crystal Oscillator and Time Chain Dividers

The basic time reference for the counter is an AT-cut series-resonant 10 MHz crystal in a simple circuit (Fig. 7)

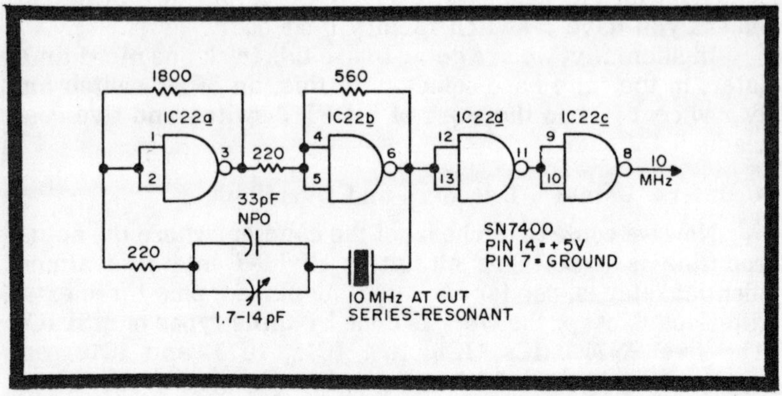

Fig. 7. The 10 MHz crystal oscillator.

using four gates from an SN7400 IC. Although a 2 MHz or even 100 kHz crystal could be substituted (with the saving of one or two SN7490 ICs in the time chain divider), it appears that 10 MHz crystals are more stable and need less temperature compensation. Besides, the 10 MHz crystal provides a better signal for zero beating against 10 MHz WWV.

The 10 MHz signal is divided down to 1 kHz, 10 Hz, and 1 Hz by seven identical SN7490 decade counters (Fig. 8). As mentioned above, a 1 MHz crystal could be used, in which case simply omit IC23 and jumper pins 11 to 14 in its place.

Incidentally, this is a good place to get accurate frequency markers for calibration; Table 1 is a partial list of frequencies available, and where to get them.

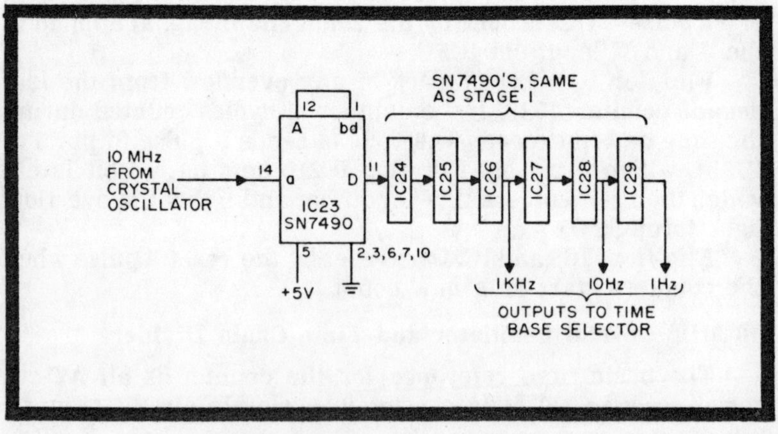

Fig. 8. Time chain dividers.

Table 1. Frequency Markers for Calibration		
Frequency	Where Obtained	
10 MHz	IC23	Pin 14
(Zero beat against WWV, 10 MHz.)		
5 MHz	IC23	Pin 12
(Zero beat against WWV, 5 and 15 MHz.)		
1 MHz	IC23	Pin 11
100 kHz	IC24	Pin 11
50 kHz	IC25	Pin 12
10 kHz	IC25	Pin 11
5 kHz	IC26	Pin 12
(Zero beat against CHU, 7335 kHz.)		

Time Base Selector

The time base selector (Fig. 9) receives 1 kHz, 10 Hz, and 1 Hz square waves from the time chain dividers, and sends two signals, depending on the position of the **Hz-kHz** switch, to the control circuits:

Switch position	CR output	S output
Hz	1 Hz	1 Hz
kHz	1 kHz	10 Hz

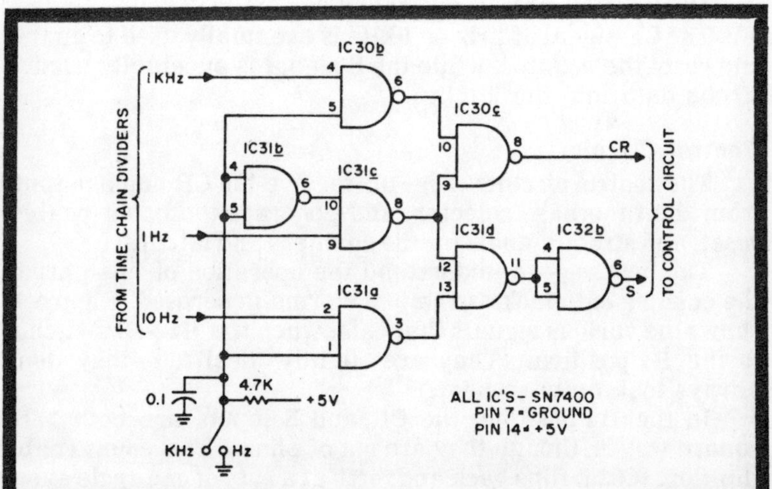

Fig. 9. Time base selector.

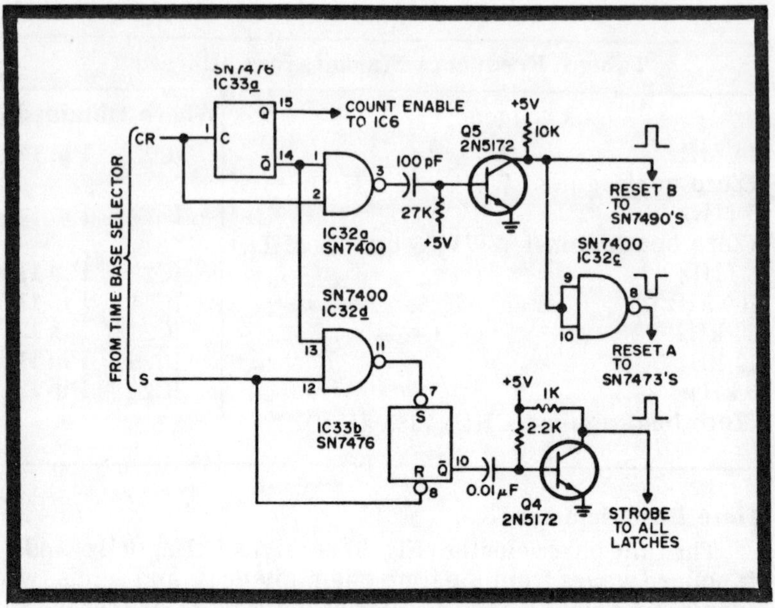

Fig. 10. Control circuits.

As mentioned earlier, the switching function could just as well be done with a DPDT switch, but this would require more wiring, and some coax cable, and might cause some problems. At the expense of two ICs (IC30 and IC31), we simplify the wiring and eliminate some headaches as well.

The CR signal at 1 Hz or 1 kHz is eventually used to enable and reset the counter, while the S-signal is eventually used to strobe data into the latches.

Control Circuits

The control circuits (Fig. 10) receive the CR and S outputs from the time base selector, and generate the count enable, reset, and strobe signals for the counters and latches.

The best way to understand the operation of this part of the counter is to look at the waveforms generated. Figure 11 shows the various signals that exist when the **Hz-kHz** switch is in the **Hz** position. (They are slightly idealized—they don't always look quite so neat.)

In the **Hz** position, the CR and S inputs are both 1 Hz square waves, though they are out of phase. The **count enable** flip-flop, IC33a, flips back and forth at a rate of one cycle every 2 sec. One complete measurement takes 2 sec, as shown. First, the count enable signal goes plus for one sec, during which time the counters count the input cycles. At the end of the first second, the count enable signal returns to zero, and a short positive strobe pulse strobes the counter outputs into the latches. A half-second later, the **reset A** and **reset B** pulses reset the decimal counters and overflow flip-flop, in preparation for the next count.

In other words, although the actual count interval takes only 1 sec, a complete measurement takes 2 sec, and the decimal readout is updated only once every 2 sec.

As shown in Fig. 12, the control circuits work quite differently when the **Hz-kHz** switch is in the **kHz** position. The count enable flip-flop, IC33a, flips back and forth and goes through a complete cycle in 2 msec. The counters count the input frequency for 1 msec, then the count enable signal returns to zero, and the counters stop. One-half millisecond later, the counters are reset back to zero, in preparation for the next count. In other words, the counters take 500 complete counts per second, each cycle of counting taking a total of 2 msec.

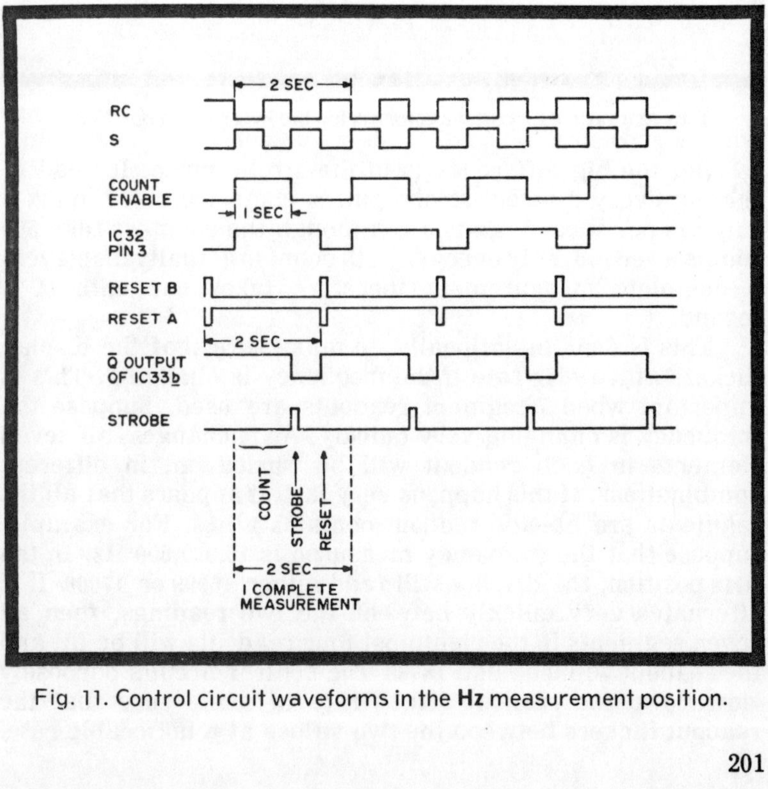

Fig. 11. Control circuit waveforms in the **Hz** measurement position.

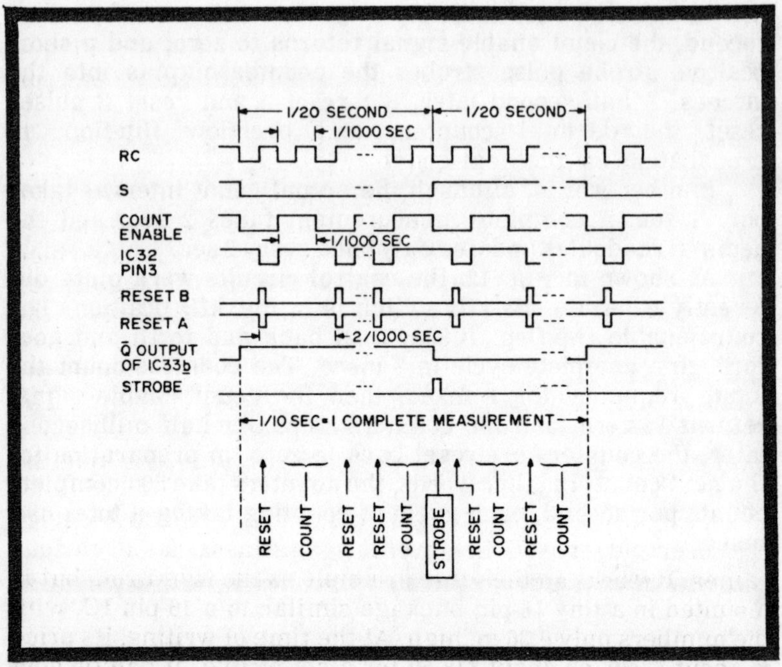

Fig. 12. Control circuit waveforms for the **kHz** switch position.

But the big difference is in the strobe pulse. Instead of coming every 2 msec, strobe pulses come 100 msec apart—only ten per second. Thus, even though the counters take 500 counts a second, only every fiftieth count is actually displayed. A complete measurement therefore takes one-tenth of a second.

This is done intentionally, to make sure that the display flickers at a 10 Hz rate if the frequency is changing. This is important when 7-segment readouts are used. Suppose the frequency is changing very quickly. As it changes, all seven elements in each readout will be turned on, in different combinations. If this happens very fast, it appears that all the segments are lit—the readout contains all 8s. For example, suppose that the frequency measured is 16,999,500 Hz. In the **kHz** position, the display will read either **16999** or **17000**. If it alternates very quickly between the two readings, then all seven segments in the rightmost four readouts will be lit, and the readout will look like **18888**. The control circuits purposely slow down the readout strobe rate to make sure that the readout flickers between the two values at a noticeable rate.

Types of Readout Devices

Several different readout devices can be used with this counter, depending on your desires and the condition of your wallet.

The PC board is designed for the SN7447 BCD-to-7-segment decoders and the RCA DR-2010 Numitron indicator tubes. The interconnections between these two units are shown in Fig. 13. The Numitron tube is a 9-pin miniature glass tube with seven thin lamp filaments, viewed from the side, which plugs into a 9-pin PC mount socket. It could be soldered directly to the board, but that's not a good idea since one of the seven filaments may burn out. Because the readout is viewed from the side, it's entirely practical to put the tube right on the PC board and mount the board horizontally in a cabinet. The decimal digits are 0.6 in. tall and quite bright. The price of one readout tube is about $4.50 at the time of writing, making the total for five tubes about $22.50.

A slightly cheaper 7-segment miniature readout device has been advertised in several magazines under various names. It works on the same principle as the Numitron, but is mounted in a tiny 16-pin package similar to a 16-pin IC, with the numbers only 0.36 in. high. At the time of writing, its price is about $3.70, or about $18.50 for a set of five. It can be connected to the SN7447 decoder as shown in Fig. 14.

Both of these readouts have one slight problem. At an applied voltage of 5V, the current per segment in the Numitron is 25 mA, and about 8 mA in the smaller readout, while the

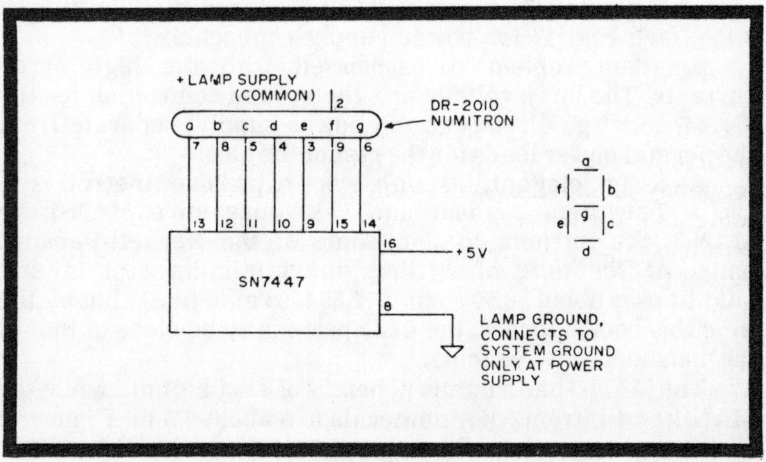

Fig. 13. Connecting the RCA Numitron (DR-2010) readout.

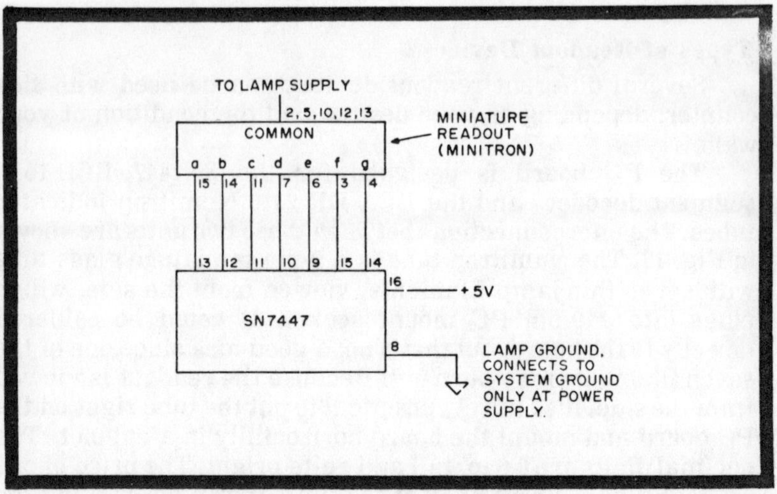

Fig. 14. Connecting the 16-pin miniature incandescent readout.

SN7447 decoder IC is rated to deliver up to 20 mA. But the surge current into an incandescent lamp filament when first turned on is about 12 times the rated current. Hence, the Numitron takes about 300 mA surge, while the miniature type takes about 100 mA. Both of these values can overload the SN7447.

One way to avoid this problem is to use the SN7447A decoder, which is rated for somewhat more current than the SN7447. Another is to use less than 5V. Some persons suggest that the lamp voltage be taken from an unregulated 3-4V source. An excellent choice is a full-wave rectified, unfiltered 3.15V (see Fig. 17 for power supply connections).

Another problem is associated with the high surge currents. The lamp voltage and the ground connection for the SN7447 must go directly to the power supply, separate from the normal power leads for the rest of the unit.

A really "elegant," though more expensive, method is to use a 7-segment readout such as Monsanto's MAN-1, or MAN-3, the Litronix 10A, or some of the Hewlett-Packard units. At the time of writing, the minimum cost in unit quantities is about $8.50 each ($42.50 for set of five); but by the time this book appears, the LED price may be close to that of the incandescent readouts.

The MAN-3 has a numeral height of about 0.1 in., while the MAN-1 and Litronix 10A numerals are about 0.3 in. Figure 15 shows the connections from the SN7447 to the MAN-1 or 10A.

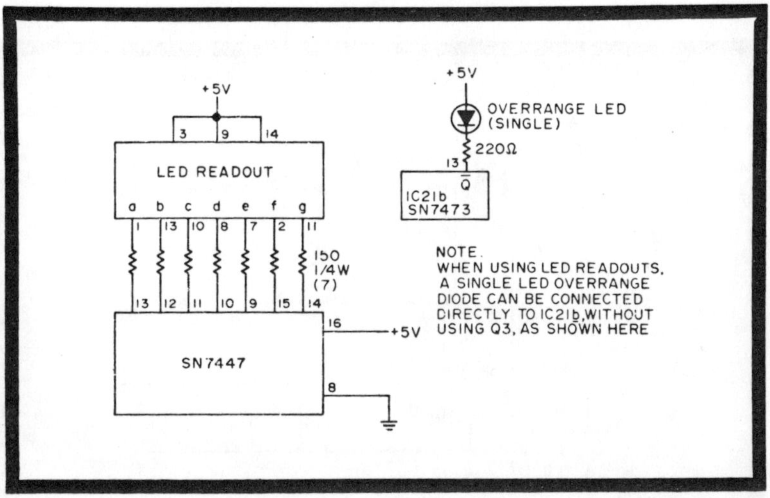

Fig. 15. Connecting the MAN-1, 10A, and LED-700 LED readouts.

Both LEDs are mounted in a miniature 14-pin package similar to a 14-pin IC.

Both the LED readout and the miniature incandescent readout are IC-like packages, which have to be viewed from above. This means that it's not practical to mount them on the same board as the rest of the counter. The best way to mount them is on a small etched or perforated board that stands up vertically right in front of the main PC board.

Power Supply

A good power supply (Figs. 17 and 18) is a very important part of any digital system. It must be well filtered and well

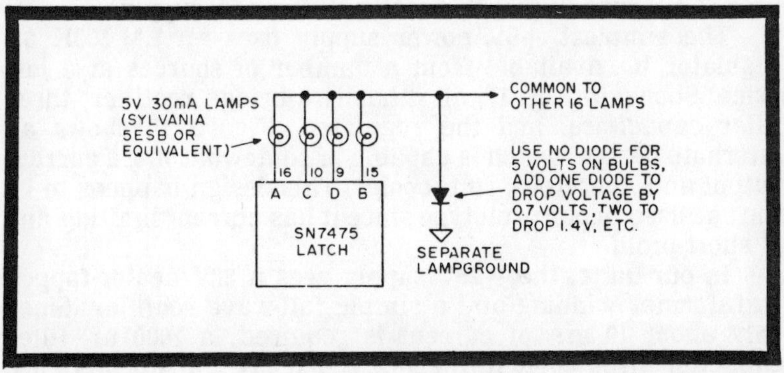

Fig. 16. Connecting four lamps to SN7475 latch, to reduce cost.

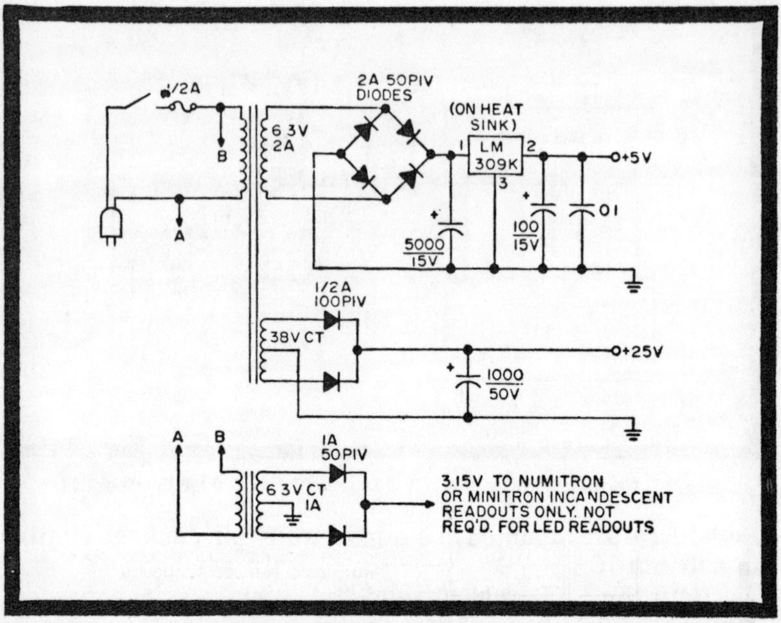

Fig. 17. Power supply.

regulated, since integrated circuits are very sensitive to noise on the power lines. Most important, the power supply voltage to the ICs should be between +4.75 and +5.25V, and must not go above 5.5V or so, to avoid damage to them.

The counter needs +5V at ±5 percent regulation at about 1.5A, +25V at about 10 mA for the FET input amplifier, and a lamp supply of about 3V and 1A if incandescent readouts are used (LED readouts work directly off the +5V supply).

The simplest +5V power supply uses an LM-309K 5V regulator IC, available from a number of sources at a low price. Shown in Fig. 17, all it has is a bridge rectifier, three filter capacitors, and the regulator. Figure 18 shows an alternate design, which is capable of somewhat more current output and, therefore, runs cooler. This design happens to be the one used in our prototype since it has current limiting and is short-proof.

In our units, the +25V supply uses a 38V center-tapped transformer winding and a simple full-wave rectifier. Since only about 10 mA of current is required, a 1000 uF filter capacitor gives enough filtering. Since 38V transformers are not very common, you may use a 16V bell transformer in a bridge rectifier circuit to give the same output voltage; the PC

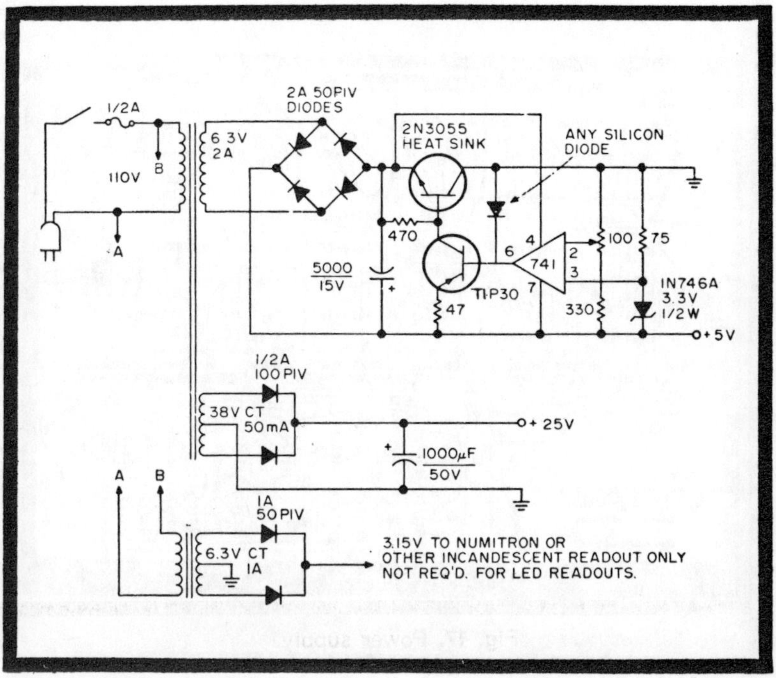

Fig. 18. Alternate power supply.

board has room for the extra two diodes in the bridge. In fact, any transformer that will provide 20-30V of dc will work. Regulation is not too important.

If you use the Numitron or Minitron incandescent readouts, the 3V lamp voltage comes from a separate 6.3V center-tapped filament transformer, as shown in Figs. 17 and 18. The two rectifier diodes mount on the main PC board. If you use the LED readouts, the transformer and diodes are not needed.

Since the TTL ICs are not only sensitive to noise, but generate sharp spikes on the 5V power line as well, adequate filtering is very important. In addition to a hefty power supply filter capacitor, there are small 0.1 and 0.01 uF disc capacitors scattered around the PC board, bypassing the +5V line to ground. These are essential to eliminate the voltage spikes as close to the ICs as possible. In fact, the two ICs in the VHF prescaler have 0.01 uF disc capacitors soldered directly under them (on the copper side of the board, connected right on the two power supply pins), just to keep out noise and prevent problems.

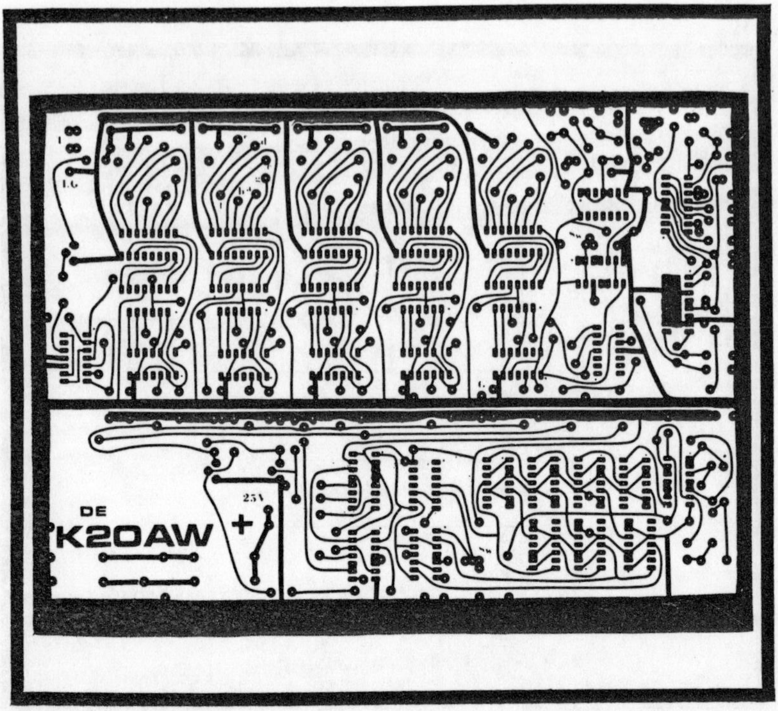

Fig. 19. Printed-circuit board layout for main board (copper side, 50 percent actual size).

Next, we will show you how the PC board is laid out, where to place the parts, and how to hook the whole thing up.

CONSTRUCTING THE COUNTER

Construction of the counter is quite straightforward, since almost all components are mounted on the main PC board shown in Fig. 19. This board is 7 x 10 in., with the readouts and the two input circuits along the front (long edge) of the card so they are convenient to the front panel of your cabinet. In order to explain where all the parts go and which part of the main board has which circuit, we've broken up the main board layout into smaller sections, shown in Figs. 20 through 26. Each of these takes one or two of the circuits shown in Figs. 3 through 10 and shows the actual layout of the parts on the board. In each drawing, a small sketch in the corner shows where to find the circuit on the main board, shown in Fig. 19.

If you want to reduce the overall size of the counter, you can cut the board in two with a fine-tooth saw, about 4 in. back from the front edge, between the two heavy copper lines running the length of the board. If you stack the rear part of the board upside-down above the front part, you can connect the two without too much trouble with short wire jumpers. But make sure to use heavy wire or braid to connect the grounds of the two boards together at the ends of the two boards.

Looking at Fig. 19, the front of the board is shown at the top, and the view is at the copper side of the board. A good way to recognize the front edge is by the five 9-pin Numitron tube sockets along the edge. In each of these figures, we are looking at the copper side of the board. The components actually mount on the other side, so we are actually looking at the bottom of the board. This is important when mounting transistors and ICs.

Input Circuit

Figure 20 shows the 0-20 MHz input circuit. Transistor Q1 is shown (bottom view) in an inset; the important thing to notice is the orientation of the little tab on the transistor case. Notice also that the ICs have to be inserted in a certain way—

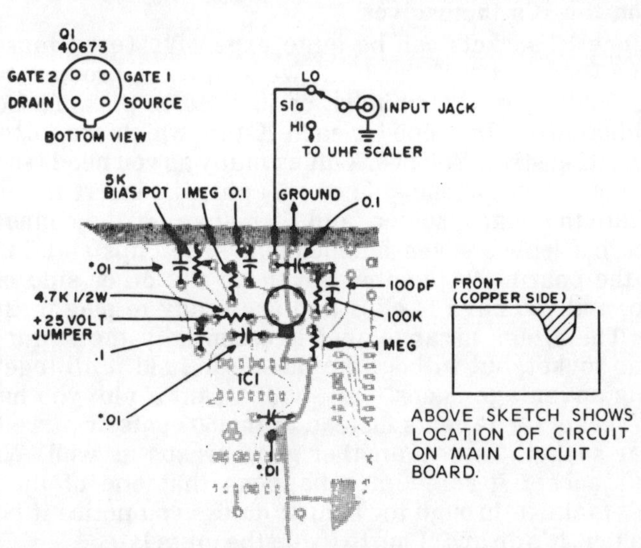

NOTES: 1. SEE FIG. 3 FOR DIAGRAM. 2. CAPACITORS MARKED ARE BYPASS CAPACITORS NOT SHOWN ON DIAGRAM. 3. OMIT S1 IF NO VHF SCALER INSTALLED.

Fig. 20. The 0-20 MHz (low-frequency) input circuit.

there is usually a little notch or depression at one end of the IC body, on the side that has pin 1. We have marked the position of this notched side with a small half-moon. Furthermore, the drawings and the board have a small dot next to pin 1. Starting at pin 1, the pins are numbered consecutively down that side, and then back along the other side. Integrated circuit IC1 is a 7-pin IC, so it has pins 1 to 7 along one side (starting from the terminal near the dot), and pins 8 through 14 back along the other side. Pin 8 is on the same end as pin 7, and pin 14 is on the same end as pin 1. Some of the ICs have 16 pins, with pins 1 to 8 along one side, and 9 through 16 along the other.

Incidentally, IC sockets of one kind or another are a must for a project like this. We don't want to scare you, but occasionally an IC is defective or you may have to troubleshoot the board, and removing an IC that has been soldered on the board is a pain in a place pills can't reach. There is another reason for sockets, and that is that some of the ICs on the board do not have all their pins connected. In order to leave more room for connections at the bottom, these pins aren't brought through the board. That means you have to remove these pins on the IC socket—it's not a good idea to start cutting pins on the ICs themselves.

Since IC sockets can be quite expensive (even more expensive than some of the ICs), we suggest a product called Molex Soldercon IC terminals, or just Molex pins. These are individual little clips, one for each IC pin, which are joined by a connecting strip. You break off as many as you need (such as a strip of seven for one side of a 14-pin IC), insert the whole strip into the board, solder, and then snap off the connecting strip. This leaves seven independent little clips standing up from the board. Put another seven on the other side of the lineup, and you have 14 clips and are ready to plug in the IC itself. The Molex arrangement is essentially the same as a regular socket, but without the plastic to hold it all together. The big advantage is cost—at about 1 cent a pin, you have a 14-pin socket for 14 cents instead of the 50 cents or more for a regular socket. There's another good reason as well. With a regular socket it sometimes happens that one of the pins doesn't make it through the board; unless you notice it before you solder, it's an awful job to fix up the mess later.

But, back to Fig. 20. The way the input jack is shown, an input signal at the jack is fed to switch S1a, and then to either the low-frequency circuit or the VHF scaler, depending on the position of the switch. Switch S1 is actually a DPDT switch, with the other pole shown in Fig. 22. If you want to simplify the switching, you might eliminate the part of S1 shown in Fig. 20 and use two separate input jacks instead. This might have the advantage of allowing you to hook up the counter to two different places, and measure the frequency at either by just flipping S1.

Prescaler Construction

Figure 21 shows the VHF scaler circuit. If you are not including the scaler in your counter, then simply omit all parts in this figure, and omit S1 as well, wiring up the input jack directly to the low-frequency input circuit instead. Otherwise, the VHF scaler is pretty much standard. Make sure here as

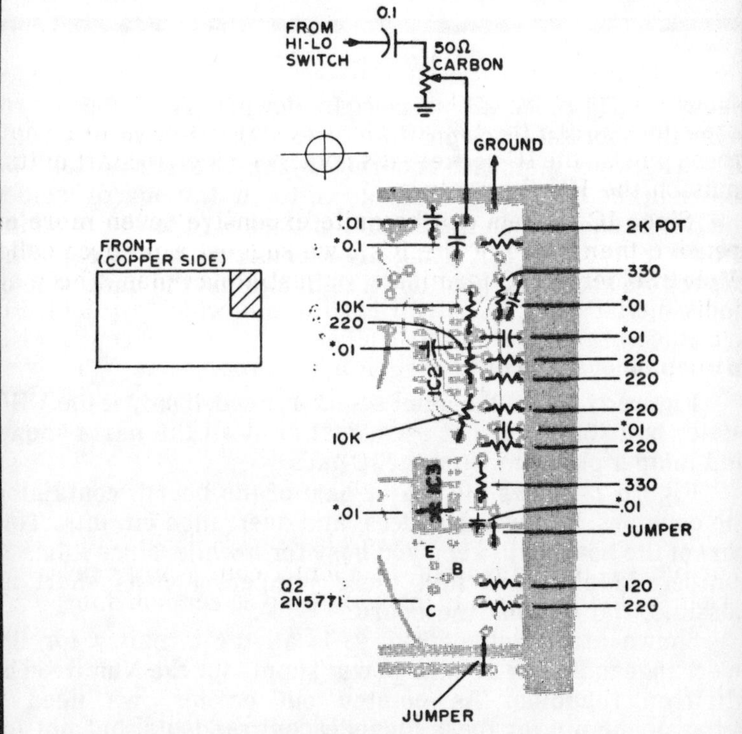

NOTES: 1. SEE FIG. 4 FOR DIAGRAM. 2. NOTE THE TWO 0.1 DISK CAPACITORS BETWEEN PINS 5 AND 12 OF IC;3. CAPACITORS MARKED ARE BYPASS CAPACITORS NOT SHOWN ON SCHEMATICS.

Fig. 21. VHF prescaler circuit.

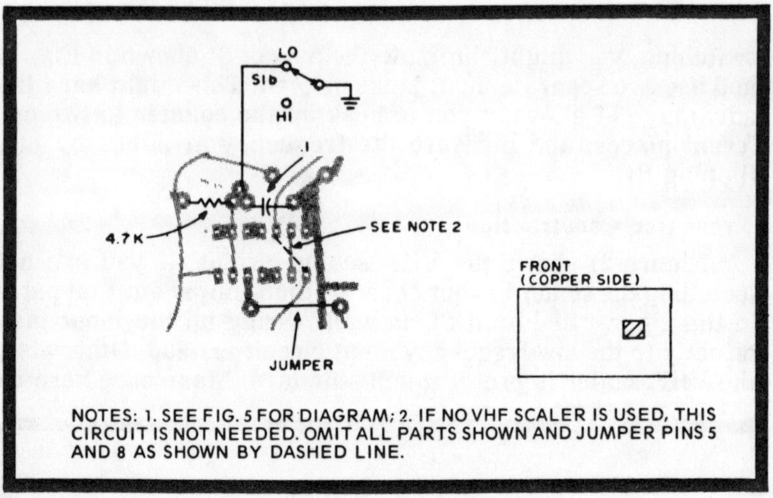

Fig. 22. Input selector circuit.

elsewhere, to place all parts right down against the board, using the shortest leads possible. Notice also the two small 0.01 uF capacitors mounted underneath the two ICs. After you solder in the IC sockets (they go on top of the board), solder the two capacitors on the bottom of the board to the two pins shown (pins 5 and 12), using the shortest leads you can. The capacitors should be bent down tight against the underside of the board.

Miscellaneous Circuits

Figure 22 shows the input selector, needed only if the VHF scaler is installed. If not used, just omit all the parts shown and jumper pins 5 and 8 on the IC pads.

Figure 23 shows the major part of the board, containing the counters, latches, decoders, and overrange circuits. This part of the board will keep you busy for a while, since it has 37 jumpers. Make sure that these jumpers are as short as possible, flat against the board.

Shown at the left of Fig. 23 is all the circuitry for the overrange indicator and the power supply for the Numitron or Minitron readouts. As pointed out earlier, we need a separate supply for these incandescent readouts, but not for LED readouts. Hence, if you decide to use LEDs, you can omit the transformer and the two diodes. You can also omit Q3 and the 10K resistor, since the LED overrange indicator can be driven directly off the output of IC21, pin 13, as shown in Fig.

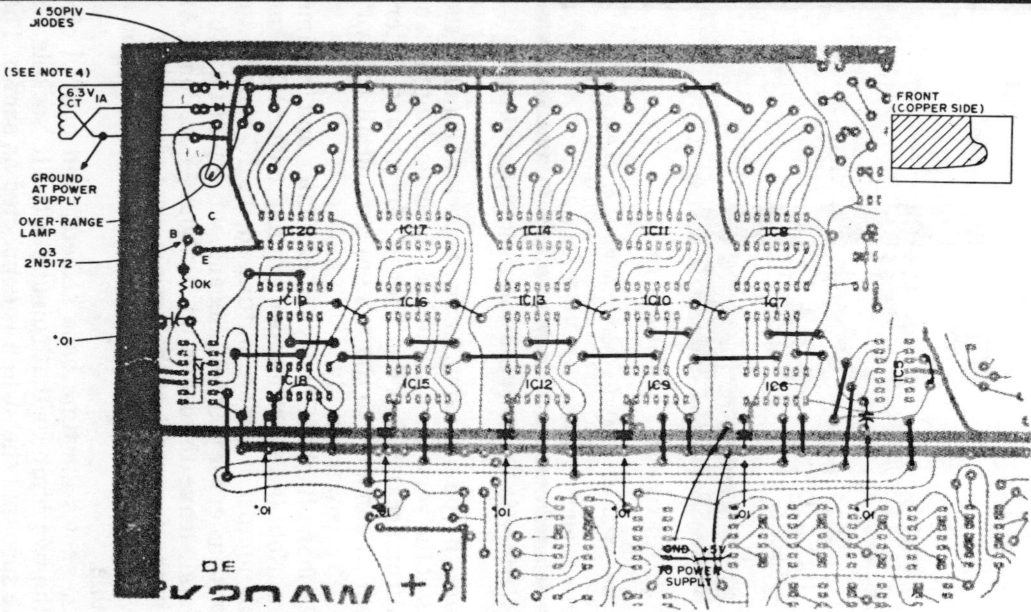

NOTES: 1. SEE FIG. 6 FOR DIAGRAM. 2. THERE ARE 37 JUMPERS SHOWN. 3. CAPACITORS MARKED ARE BYPASS CAPACITORS NOT SHOWN ON DIAGRAMS. 4. Q3, TRANSFORMER, DIODES AND ASSOCIATED PARTS NOT NEEDED FOR LED READOUT.

Fig. 23. Counters, latches, decoders, and overrange circuitry.

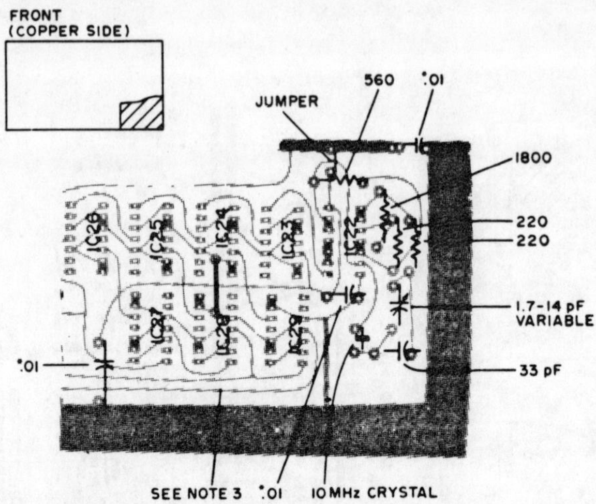

NOTES: 1. SEE FIGURES 7 AND 8 FOR DIAGRAM: 2. CAPACITORS MARKED ARE BYPASS CAPACITORS NOT SHOWN ON DIAGRAM. 3. NOTE THE JUMPER RUNNING UNDER IC24 AND IC28.

Fig. 24. 10 MHz crystal oscillator and time chain divider circuits.

28. Incidentally, if you are using LEDs, you can also omit the four jumpers at the very front of the board, connecting pin 2 of all the Numitron sockets, but don't forget the ground to terminal LG, which connects to the system ground all the way back at the +5V power supply.

Figure 24 shows the 10 MHz crystal oscillator and the time base dividers. Note the short jumper running under IC24 and IC28. Put this jumper in before putting in the IC sockets, to make the work easier for yourself. Second, when putting in the 10 MHz crystal, leave the leads on it about ½ in. long. As pointed out in the first part of the article, someday you may want to place the crystal in an oven, and at that time you'll appreciate this.

Figure 25 shows the time base selector and control circuits. Note the two jumpers that run under the IC sockets, as before. Make sure that the switch is connected as shown. The lead to the switch can be a single unshielded lead, running either under or over the board, up to the front panel. Virtually any silicon NPN transistors can be used for Q3, Q4, and Q5, as long as they have reasonably good gain, but we suggest the 2N5172, since it is very cheap.

Figure 26 shows the +25V supply, mounted in a corner of the main board. As shown, the circuit is designed for a 38V center-tapped transformer, and uses a full-wave rectifier. If you have a 16-20V transformer, add two diodes as shown and eliminate the centertap. This makes the circuit a full-wave bridge. Watch the polarity on the electrolytic, and use a 50V unit—not a 25V unit!

As part of the +25V supply, you will need a long jumper from the 25V terminal over to the 4.7K resistor near Q1, shown at the left of Fig. 20.

Readout Board

As mentioned earlier, the easiest readout to use is the Numitron...just install 9-pin miniature tube sockets on the board and plug in the readouts. In one of our prototypes, however, we used LEDs because of their good looks. Figure 27 shows the LED readout board for mounting them. In addition to Molex pins or 14-pin IC sockets, you will need thirty-five 150-ohm resistors for current limiting. The overrange indicator also mounts on this board. The board has room for two overrange indicators, one above the other, to make a sort of colon (:) arrangement, but only one is really needed. For LED

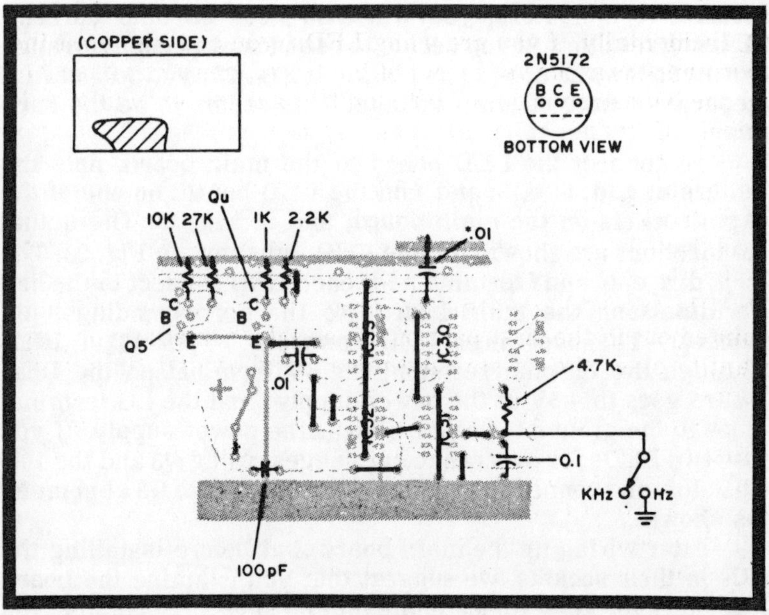

Fig. 25. Time base selector and control circuits.

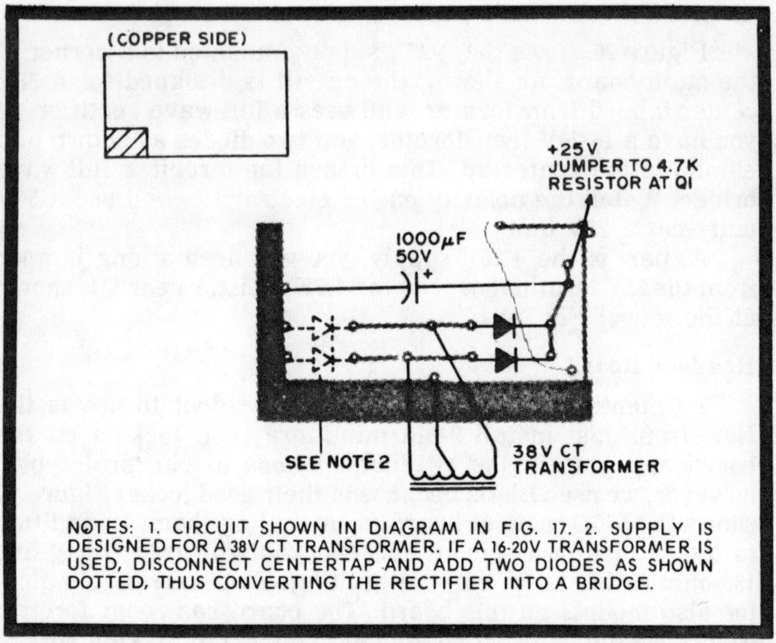

Fig. 26. The +25V power supply.

overrange indicators, you will also need 220-ohm current-limiting resistors. If you decide to use incandescent overrange bulbs, follow the hookup as shown in Fig. 23; you will need a separate source of lamp voltage. This seems to be the hard way.

To connect the LED board to the main board, note the letters **c, g, d, a, e, b,** and **f** on the LED board, on one of the 9-pin sockets on the main board, and in Fig. 15. The actual connections are shown, for one LED indicator, in Fig. 28. The c, g, d, a, e, b, and f terminals for each LED connect to the like terminals on the main board (to the corresponding 9-pin socket). The overrange LED connects to pin 13 of IC21, through the 220-ohm resistor; the 5V terminal on the LED board goes to +5V at the power supply; and the LG terminal goes to the ground terminal, also at the power supply. If you use two LEDs for overrange, we suggest using Q3 and the 10K resistor, and connecting the 220-ohm resistors to Q3 at point X, as shown.

After wiring up the main board, but before installing the ICs in their sockets, we suggest that you examine the board carefully for good solder joints and no shorts. Hopefully, you

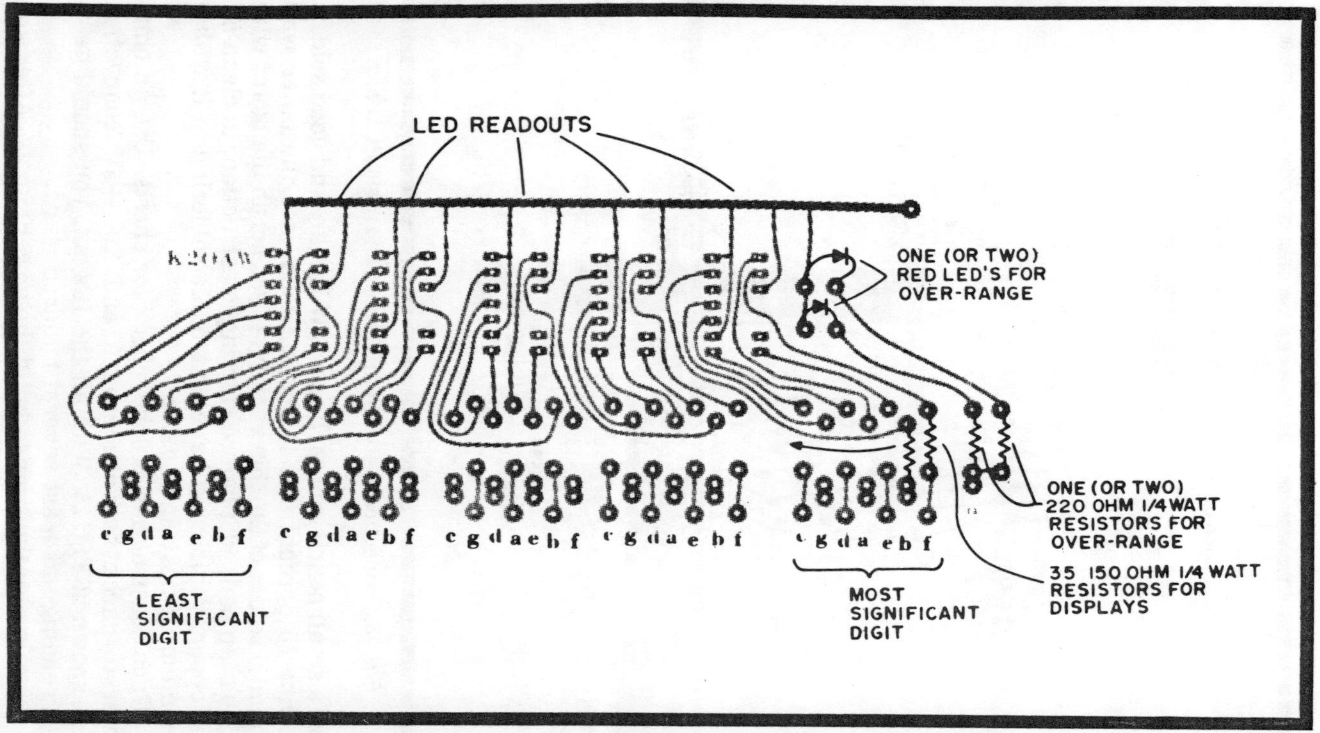

Fig. 27. Rear view (copper side) of board for mounting MAN-1, 10A, or LED-700 readouts.

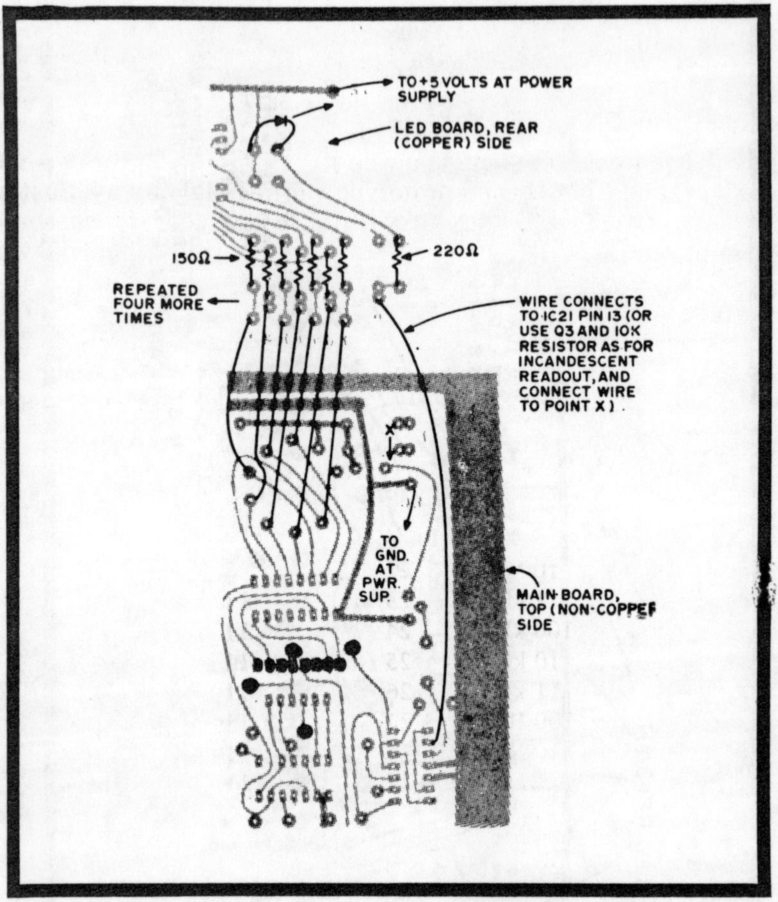

Fig. 28. Connections between LED board and main PC board.

used a small pencil iron such as a 30W Ungar, and good solder. We like 60 percent tin, 40 percent lead (not the other way around!) solder of the thin variety. If you scrub the board with Brillo soap pads just before starting construction, to clean off the copper, you should have no trouble soldering to it with a minimum of heat and solder.

Once construction is done, check for shorts. Use an ohmmeter to check between ground and the +5V connection point shown in Fig. 23. With all the ICs out, you should get a high reading—at least several hundred ohms.

Now is also the time to build the power supply. After you finish it, make sure to test it and, if needed, adjust the output voltage to +5V before connecting to the main board. An overvoltage can ruin a lot of expensive ICs.

Once everything is hooked up, take the big step and turn on the main power. If all the indicators read 00000, you will probably breathe a sigh of relief—but all is not done yet. With luck, your counter should now be working. Except for two bad connections, our second prototype worked right away. Just in case your counter doesn't work and you need to do some troubleshooting, here is a suggested way of going about it.

First, check the 10 MHz oscillator and time chain dividers. Put a meter (20,000 ohm-per-volt VOM) at IC29, pin 11. The meter should swing up and down at the rate of one cycle per second. If so, fine; if not, use a scope to trace the signals through the circuit. Check the points in Table 2 for the signals shown.

Table 2. Oscillator and Divider Signals

Signal	IC	Pin
10 MHz	22	8
1 MHz	23	11
100 kHz	24	11
10 kHz	25	11
1 kHz	26	11
100 Hz	27	11
10 Hz	28	11
1 Hz	29	11

If no 100 MHz signal appears at IC22, interchange the SN7400N IC with one of the others in the counter. If the signal seems to go down the line but stops at some point in the divider chain, interchange the affected SN7490N with another. In this way, you should be able to get the oscillator and divider chain working.

Next step, if everything is okay so far, is to check whether the indicators light. If not, check the power connections to the indicators, and the ground and +5V connections to the SN7447N decoder-drivers. If some indicators light and others don't, interchange ICs and indicators (with the power off!) to narrow down the cause. If interchanging the ICs doesn't do it, check your connections.

Once the indicators light, they will probably read all zeroes. Place the Hz-kHz switch in the Hz position, and turn off the power for about 5 sec. When you turn the power back on, the indicators should come back with some really crazy readings, or perhaps even be totally dark. It should take about 2 sec for them to reset back to 0. This is because the flip-flops in the counters and latches come back in random states, and it normally takes 2 sec to get everything reset back to 0. If this occurs as described, fine. Otherwise, check the control circuits for waveforms similar to those in Figs. 11 and 12, depending on the position of the **Hz-kHz** switch. Unless you have a really good scope, you may not be able to see the thin pulses on the reset and strobe lines.

CHECKOUT AND OPERATION

Once everything seems to work okay, remove IC4, and connect a short wire to pin 8 of the socket. Place the **Hz-kHz** switch in the **kHz** position, and use the wire to trace the signals listed in Table 2, starting at 1 kHz and going up in frequency. The 1 kHz signal should result in a reading of **00001** on the displays, the 10 kHz should result in **00010**, and so on. Go through all five counters like this, checking for a **1** in each place. If everything works fine at lower frequencies but not at the higher frequencies, check the SN7490N counter at the rightmost position that doesn't work by interchanging it with another to its right. Don't go back and swap ICs in the time chain dividers, though—leave well enough alone. If most of the readings here are okay, but one of the digits in the center does not seem to work, check the counter, latch, and decoder as well. If you use any readout other than the Numitron, and get some strange readings, check the wiring between the readouts and the main board for a mixup.

Once you have gone this far, you are pretty close to being home. Try both input circuits. If one input works and the other doesn't, that should narrow down the trouble. If neither works, check the circuitry around IC4. Check also the wiring to S1—you may have the wiring mixed up so that the input of one of the two input circuits is connected at the same time as the output of the other. You won't get too many signals through that way.

Once everything is working, readjust the 2K and 5K pots for best sensitivity. Don't be surprised if the VHF input shows some reading even with no input connected. This indicates some oscillation in the input amplifier, and you can easily eliminate it by slightly moving the setting of the 2K pot. We've seen even some fairly expensive commercial counters do this, and the oscillation usually gets swamped out when an actual signal is connected.

The inputs are actually quite sensitive. We have a local radio station on 1310 kHz—a mile or so away—and we often notice the counter with a **01310** indication when connected into a circuit that has its power turned off. If you forget to connect the ground to the input of the counter, you may often see a **00060** reading in the **Hz** position—that's just plain hum. It's nice to know that your local electric company keeps its frequency on the nose.

Speaking of frequency, there is one more adjustment to be done, and that is zeroing of the 10 MHz oscillator against WWV. This is quite simple, as the counter puts out a nice 10 MHz signal with the case off. If you find that the crystal won't quite go to 10 MHz, you can change the value of the 33 pF capacitor in parallel with the trimmer capacitor as needed. If you can't receive WWV, then you might connect another (calibrated) counter to the 10 MHz signal and adjust that for the right frequency. Heathkit has another trick in their instruction manual for their counter, and that is that you place an ordinary transistor AM radio near the counter, and tune to a broadcast station. Since the counter puts out strong signals at 10 and 20 kHz as well, the harmonics of these will beat with the station. Since the FCC requires commercial stations to be close to their assigned frequency, which is always on a multiple of 10 kHz, you can zero beat the counter against the station.

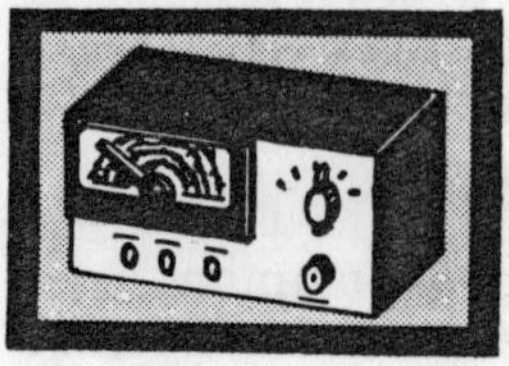

Digital Readout For Signal Generators

Richard Factor

A large percentage of transceivers and signal generators can be fitted with a digital dial with great accuracy, and at a very reasonable cost. Little modification of your gear is required, and what is required will not affect the resale value, since it is invisible and easily removed.

As an extra added plus for the homebrewer, the digital dial makes unnecessary the greatest hate object of the electronic purist: the tuning dial, with its attendant difficulties of getting linearity and accuracy—and, above all—the difficulty of inscribing, decaling, engraving, or calligraphing the dial in a neat and readable manner.

The device to be described is an adaptation of the basic electronic counter circuit. It takes the VFO signal of your generator or transmitter and tells you what it is...accurately—not to the nearest kilohertz, but down to the nearest **10 Hz**, almost always, with no precautions or requirements other than a periodic check to zero beat the oscillator with WWV. If it is installed in a transmitter, it can measure received signal frequency by zero beating the transmitter with the receiver. It can also measure the frequency shift of a RTTY signal by zero beating the mark and space frequencies and subtracting the two readings.

PLANNING FOR THE DIGITAL DIAL

You know frequency counters are expensive, and you're scared away by the price. Right? You shouldn't be...I mentioned that IC and readout prices are declining. Below is a comparison based on the advertised prices of the major components involved. I haven't included resistors and small components, because you no doubt already have them. Even if you don't, the total should be less than $10. It is assumed that your rig has high-voltage and filament supplies so that no power transformer must be bought.

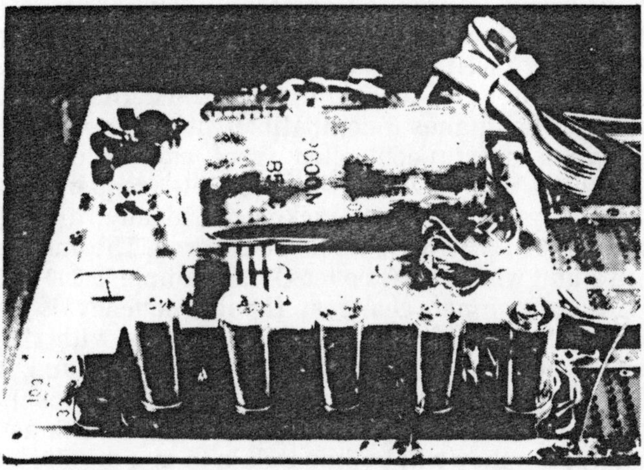

Front view of digital dial, showing sandwich construction.

Type	Quantity	Total
SN7490	10	$20
SN7475	5	$10
SN7441AN	4	$12
B5750 (Nixie)	5	$25
1 MHz xtal	1	$ 5
		$72

This is effectively for the "worst case," i.e., no external oscillator, using a 1 MHz crystal, and using a 5-digit readout with storage. By eliminating the 10 Hz resolution, using an existing 100 kHz calibrator, and deleting the most significant digit, the price bcomes drastically lower.

Type	Quantity	Total
SN7490	7	$14
SN7441AN	3	$ 9
B5750	3	$15
		$38

Since dial-drive mechanisms with less than one-tenth the accuracy cost over $30, the price is not at all out of line. Incidentally, the prices quoted above are the highest you should have to pay. Due to a current oversupply in the IC industry, substantial discounts may be obtainable, and there is little doubt that prices will drop substantially between the time these words were written and the time this appears in print.

Now that you have decided to build the digital dial, let's see if its circuitry makes it compatible with your rig. The rig should have a crystal-controlled front end or conversion oscillator and a VFO covering a reasonable frequency range without odd-number kilohertz tacked on. Ideally, a range of from, say, 5.000 to 5.500 MHz would be covered. The unit can be used with a unit whose VFO covers, for example, 5.300 to 5.800 MHz, with only minor changes. Input frequency is unimportant—anything up to 15 or 20 MHz is okay with the ICs specified. What is important is that there be no odd numbers at the end. A VFO covering 5.455 to 5.955 would be unacceptable. Of course, the counter will measure such frequencies as well as any other, but they will be impossible to mentally relate to the operating frequency. Another requirement is that the frequency-mixing scheme be either additive or subtractive, but not a combination of the two. A quick glance at your instruction manual will tell you the exact VFO frequency and mixing scheme. If your rig tunes in the same direction on all bands, you should have no problem with the additive-subtractive question. If your rig does not fit the above criteria, I'll have a few comments on possible remedies, untried, but theoretically sound, at the conclusion of the article.

The digital dial described herein is being used with my HX500 transmitter, whose VFO covers 3.9-4.4 MHz, and which employs subtractive mixing. However, the principles and circuitry can be used with only minor modification in any transmitter or receiver fitting the above criteria.

THE CIRCUIT

The first step in going digital is to modify your rig. The digital dial needs +5V dc at about 500 mA, B+ at about 15 mA, and 1V of signal at the VFO frequency. To get the 5V, tap the filament supply (6.3 or, preferably, 12.6V), and connect it to the bridge rectifier shown in Fig. 1. If your filament supply is 6.3V, it may be necessary to connect another 6.3V, 500 mA transformer in series with the filament supply. Nixie tubes "like" 170V at about 3 mA across them, but since they tend to act as voltage regulators, current limiting is necessary. Measure your B+ supply, subtract 190V, and calculate a resistor that will give 3 mA per Nizie with this voltage across it. Be sure to make a power calculation also, as a 2W resistor

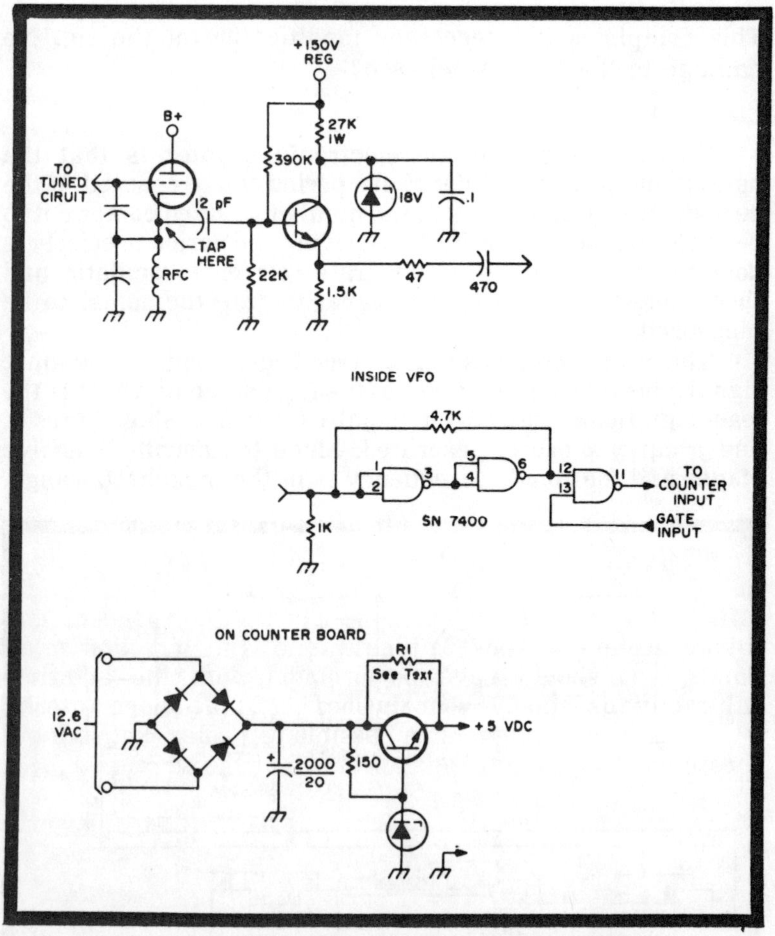

Fig. 1. V_{fo} isolation, input circuit, and power supply.

required. Small series resistors are connected to each Nixie to equalize current; the resistor just calculated goes to the junction of the series resistors.

The purpose of the VFO buffer is to make sure that the VFO in the rig is not disturbed. An emitter follower is ideal for this application. The component values shown are for a supply of 150V, as this is the most commonly used. No trouble should be encountered if the pickoff capacitor (12 pF in the schematic of Fig. 2) is at least an order of magnitude (10 times) smaller than the capacitor from the cathode to ground. The actual capacity will be much smaller, since the pickoff capacitor is in series with the transistor base, but it doesn't hurt to be safe.

This completes the necessary modification of the rig. No damage to the front panel—see?

Circuit Description

The theory behind the electronic counter is that the number of cycles of the signal in a period can be counted. If the period, for example, is 1 sec, the number counted comes out to be cycles per second, or hertz. Deriving the period precisely is done by counting down an accurate reference oscillator and then using the counted-down signal to gate the signal to be measured.

The circuit works as follows (see Figs. 3 and 4). The input signal goes into a divider chain, the first stage of which is the least significant digit. After 10 pulses, the first stage is reset, and a carry pulse is generated. Since the counter has five stages and the counted frequency is in the megahertz range,

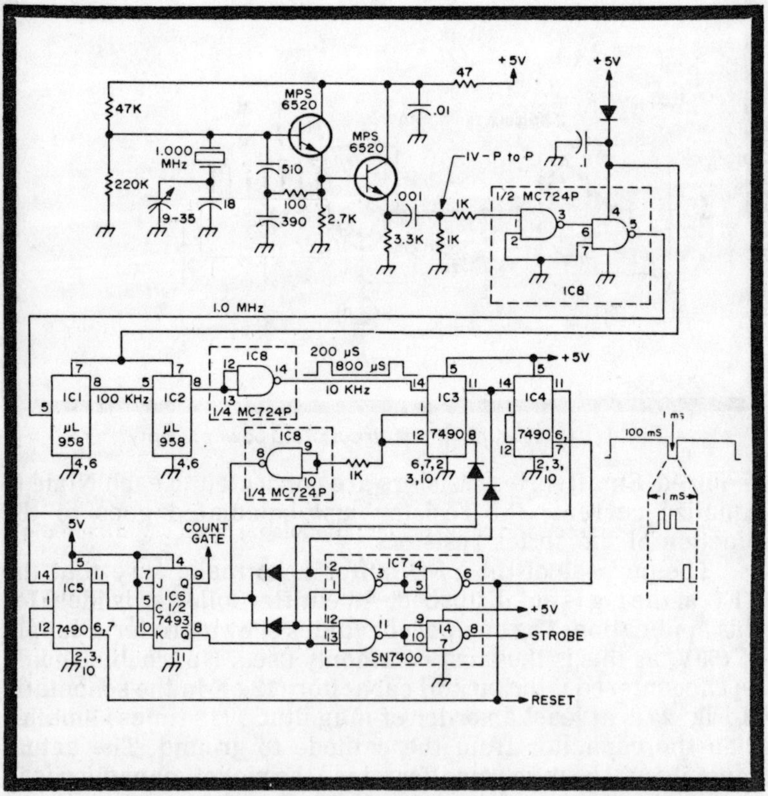

Fig. 2. Frequency standard oscillator and counter control circuitry with timing diagrams.

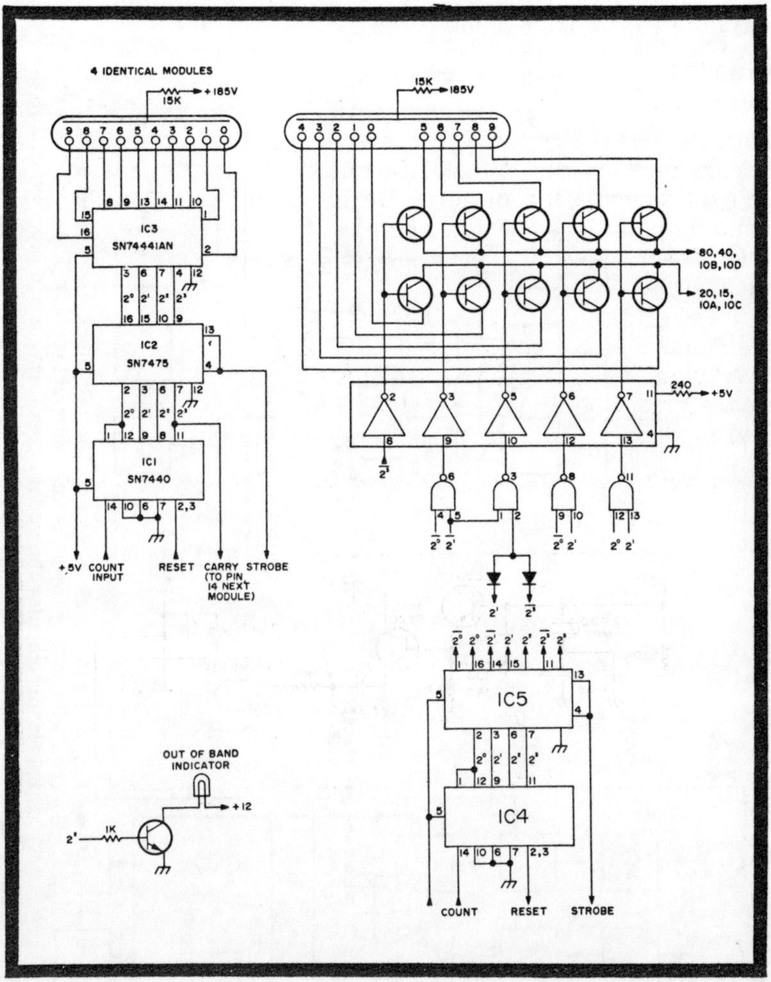

Fig. 3. Digital counter and Nixie driver; module for least significant digit is shown.

you can see that the counter will overflow several times during each count interval. However, we know what the first digit will be, so there is no point wasting a counting stage on it. All the stages are identical except for the most significant digit, which uses discrete components since special decoding is required.

After the counters have counted, the SN7441AN decoder-drivers ground the appropriate Nixie cathodes, and the corresponding numbers light up. Between the counters and the

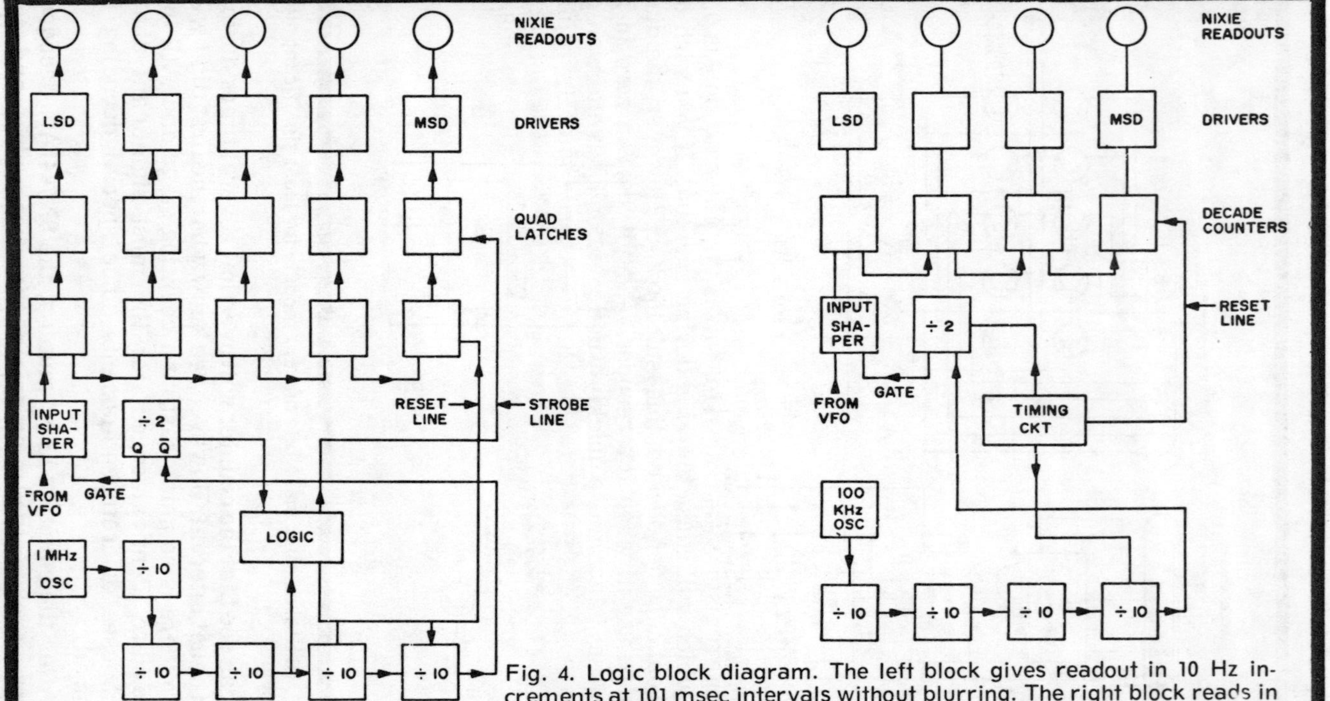

Fig. 4. Logic block diagram. The left block gives readout in 10 Hz increments at 101 msec intervals without blurring. The right block reads in 100 Hz increments; the update is adjustable and blurring occurs during 100 msec count.

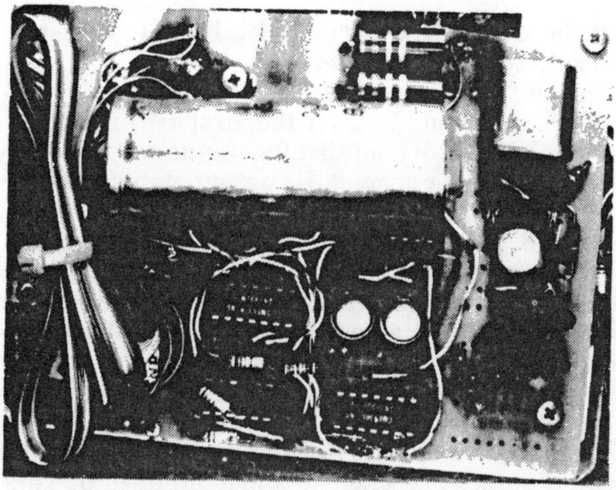

Top view of power supply and timing section.

decoder-drivers, one additional stage is necessary, a buffer storage register. The reason for this is that the gate time is 0.1 sec, which will cause blurring during counting. If you wish to read out once per second, this isn't too bad. However, the dial is more useful when it's responsive, and you can zero in a frequency much more quickly when you don't have to wait a full second to see where the dial turn took you. By using storage registers and a few logic gates, it is possible to make ten 100 msec measurements in 10.1 seconds with no digit blurring. How this is done brings us to the timing circuitry.

First, an accurate frequency must be generated. The circuit shown in Fig. 2 is a stable oscillator designed for 32 pF crystals. A trimmer is included for fine frequency adjustment. There is no reason why the 1 MHz output of another oscillator can't be used. Approximately 1V p-p is required. You can also use a 100 kHz oscillator and delete one divider stage. The MC724P quad 2-input gate shapes the signal into a square wave suitable for triggering the dividers.

The first two dividers I used in my unit are Fairchild uL958s. I used them because I had them lying around, and it is okay to use SN7490s here. A section of IC8 is used as a buffer to make the two IC families compatible. If you use 7490s, delete this gate.

ICs 3, 4, and 5 are the remaining dividers. At time zero, let us suppose that IC6, the gate flip-flop, is reset (gate at logic 0). The carry pulse from the last divider brings the gate "high," starting the count. One carry pulse later (100 msec), IC6 is again reset, stopping the count.

Here the fun begins. The IC6 Q-output, being high, sets half of the AND condition on the first two sections of IC7. Meanwhile, IC3 is still counting from binary 0 to binary 9 at a rate of 1000 times per second. Its output states are decoded by the remaining inputs to IC7 in the following sequence: At binary 4 and 6, pins 12 and 13 are both high, causing pin 11 to go low, pin 8 to go high, and thus strobing data into the buffer register. The strobe is performed twice, because doing so eliminates the need for an additional decode. The same data is strobed each time.

At binary 8, pins 2 and 1 of IC7 are both high, causing pin 3 to go low and pin 6 to go high, thus resetting the counter and resetting the last two 7490s in the dividing chain. However, these two dividers are reset to 9 instead of to 0. This means that just 0.1 msec later, when IC3 carries, IC4 and IC5 will also, starting the count again. Note that the total time between the end of the first count and the beginning of the next is just 1 msec, allowing almost 10 measurements per second.

Backwards or Forwards?

If your rig is of the subtractive-mixing type, you are probably wondering how the readout frequency corresponds to the output frequency of the rig. It doesn't—everything is upside-down. Of course, it will be accurate on one frequency in the middle of the band, but that's little comfort. Despair not.

Top view of counting circuit.

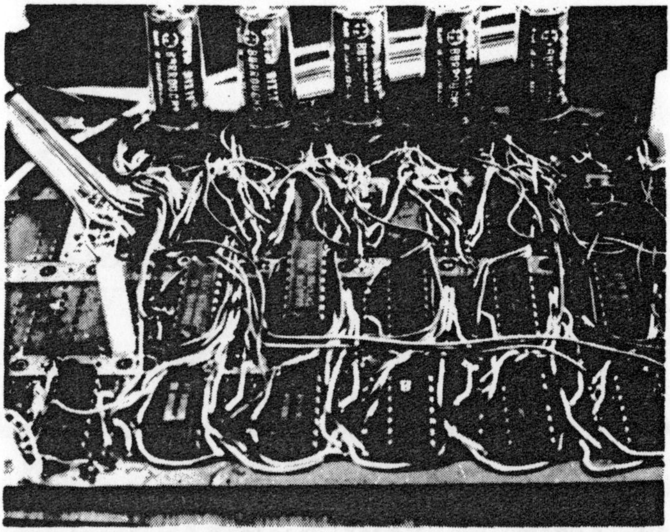

Rear view of counting circuit.

Instead, connect the Nixies to the counter **backwards**. (Switch 0 for 9, 1 for 8, 2 for 7, 3 for 6, and 4 for 5.) Through the magic of mathematics, for every 10 Hz increment of the VFO frequency, there is a 10 Hz decrement of the readout frequency. You can now see why I said that the rig can be additive or subtractive, but not both...unless you happen to have a 2-position, 50-pole switch around.

Most Significant Digit

The one remaining readout problem is that of the most significant digit. If you want the digital dial to be accurate on all bands it must read: (3) 745.92 kHz on 80 meters, and (7) 245.92 kHz on 40 meters. (The 3 and 7 are "understood.") One way this can be done is to have different drivers for the Nixie, and to ground the appropriate one with an extra wafer on the bandswitch (see Fig. 5). By driving the bases in parallel and grounding the emitters of the appropriate group, a cheap and dirty "adder" is constructed.

If your VFO starts from some frequency other than an even megahertz, the decoder must be designed to match. If the VFO is, say, 3.9 to 4.4 MHz (as is the HX500), where 3.9 corresponds to the top of the band and 4.4 to the bottom, then the decoding is as follows: 4.399.99 to 4.3000.00 is **000.00** to **099.99** kHz on the dial. Thus, 3 must be decoded as 0, 2 as 1, 0 as 3, and

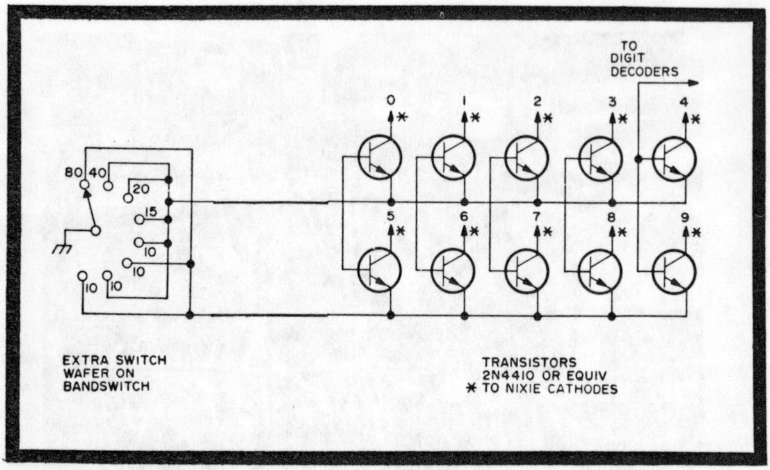

Fig. 5. Using a wafer switch to ground the appropriate Nixie drive for "adder" function.

9 as 4. To do this, the gate that decodes a 3 binary state should drive the 0 (and 5) digit line, etc. The decoding section is fairly simple, since only 5 states must be decoded, compared to the 10 states decoded by the SN7441AN. Figure 6 and Table 1 give the general decoding logic for all possible binary-coded-decimal (BCD) states. Considerable simplification is possible if you have a good understanding of digital logic. For example, the schematic of my unit shows the decoding function performed by one 3-input gate equivalent, three 2-input gates, and one 1-input gate.

If you wish, the unused states of the counter can be used to give an out-of-band indication (Fig. 7). None of the N outputs

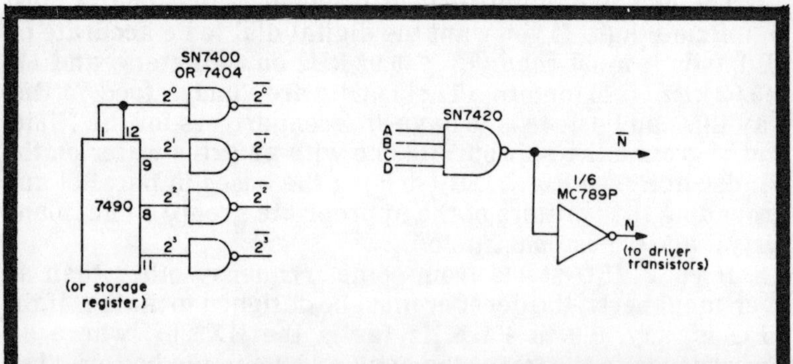

Fig. 6. General decoding scheme for all BCD states.

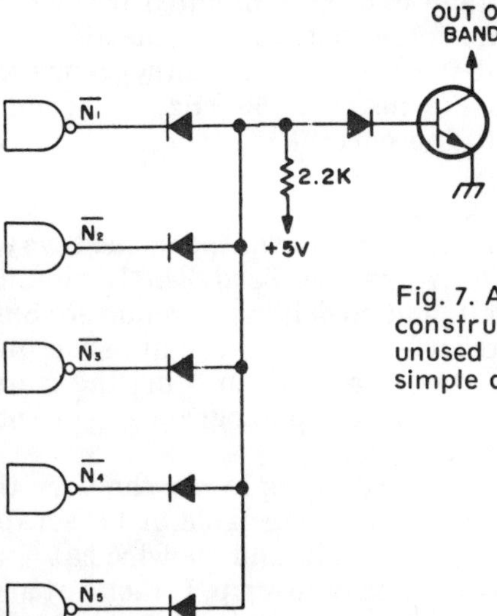

Fig. 7. An out-of-band indicator is constructed by ANDing the unused **N** states and feeding a simple driver.

will be low if none of the proper states is decoded. This allows the transistor to be saturated. You can use it to light a light, ring a bell, disable the VOX circuit, or start a tape recording with hosannas to the FCC!

"El Cheapo" Model

Thus far, my comments have been about a "deluxe" counter with storage and 10 Hz readout. If you wish to build a more austere model (read "cheap"), there are ways to cut down on the cost. The best is to make a 4-digit model with a 10 msec time base. Since the count interval is so short, you can sample once every 0.1 sec, count for 0.01 sec, perform housekeeping functions in 2 msec, and have a display visible for almost 90 percent of the time, without using storage registers. One Nixie stage is eliminated, and IC6 is unnecessary.

To make the stripped-down version, substitute the circuitry of Fig. 8 for the bottom half of Fig. 2. It works as follows: As the dividers count up from 0, the diodes decode a count of 88 and generate a 1 msec reset pulse. At the transition to 90, the count gate decodes a binary 9 on the most significant digit and goes high for 10 msec, forming the gate interval. Thus, the SN7490s perform the essential timing functions.

An additional Nixie stage can be eliminated by deleting the most significant digit from the count chain. This also saves the trouble of figuring out the decoding circuitry. Since any VFO will be accurate to the nearest 100 kHz, a second of mental work will compute the correct number.

Construction

Unless you want to make a PC board, the easiest way to build the unit is to use a "universal" IC card. The ICs and the Nixies can be mounted on it, and no drilling is required. Since most of the wiring is repetitive, it can be accomplished quite rapidly. The counter board will take about 4 hr, the timing board about 2 hr. Looking at the photographs, you may disbelieve me, but it goes very fast.

One suggestion to speed things up is to use the wire that can be found in multiconductor telephone cable. A 4 ft scrap of this stuff has 200 ft of hookup wire in it, and the wire has a soft plastic insulation, which is very easy to strip. It melts at a low temperature, but since you're using a low-power soldering iron, there should be no problem.

The crystal oscillator is quite stable, but it can obviously be imporved by installing the crystal in an oven. The TTL circuits used in this project are much better than RTL with regard to noise immunity. A significant amount of rf can be floating around before erratic operation occurs. However, it is

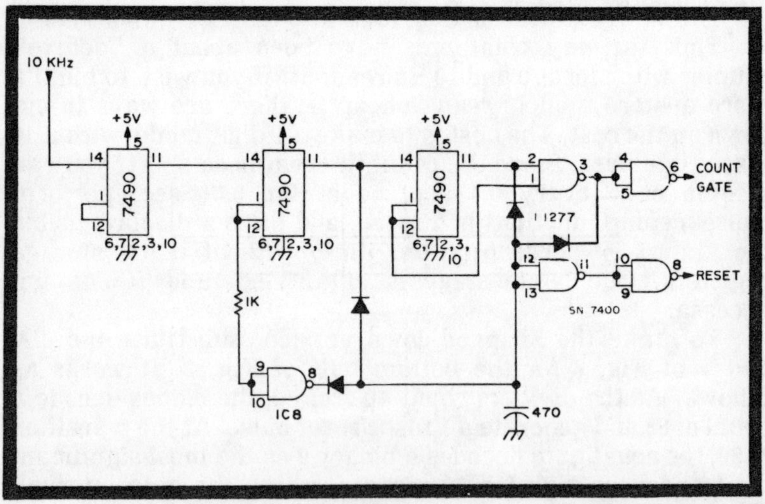

Fig. 8. By substituting this circuit for the bottom half of Fig. 2, a cheaper but less accurate digital counter results.

The resistor shown bridging the transistor bypasses some of the current without, theoretically, greatly affecting the

I used some nondescript power transistor for the series regulator. Anything that can handle an amp and 5W of dissipation should be sufficient.

recommended that the unit be carefully bypassed and shielded; first, because often there is a considerable amount of rf floating around, and, second because the pulses in the digital circuitry are very fast and can radiate noise to the receiver.

Mechanically, my unit was constructed on two boards, which were then sandwiched together. This saved about 10 sq in. There is no reason that the unit can't be built on one board. Final dimensions exclusive of case were 5 x 4½ x 2¼ in. Something this size could probably be installed inside the cabinet of the rig. If you wish, the Nixies and their drivers can be placed on a separate board, which should easily fit in the space taken by the typical dial drum or scale assembly.

The unit requires 5V and some convenient high voltage. I described the B+ supply earlier. The 5V supply is worth a few words. Regulation is not critical, but if the supply has lots of hum or noise, filter it out before it gets to the oscillator. The oscillator only takes 5 mA or so, and an RC filter should be sufficient.

Table 1. Connection Chart for all Binary Combinations

		CONNECT			
N=	BINARY	A	B	C	D
0	0000	$\overline{2^0}$	$\overline{2^1}$	$\overline{2^2}$	$\overline{2^3}$
1	1000	2^0	$\overline{2^1}$	$\overline{2^2}$	$\overline{2^3}$
2	0100	$\overline{2^0}$	2^1	$\overline{2^2}$	$\overline{2^3}$
3	1100	2^0	2^1	$\overline{2^2}$	$\overline{2^3}$
4	0010	$\overline{2^0}$	$\overline{2^1}$	2^2	$\overline{2^3}$
5	1010	2^0	$\overline{2^1}$	2^2	$\overline{2^3}$
6	0110	$\overline{2^0}$	2^1	2^2	$\overline{2^3}$
7	1110	2^0	2^1	2^2	$\overline{2^3}$
8	0001	$\overline{2^0}$	$\overline{2^1}$	$\overline{2^2}$	2^3
9	1001	2^0	$\overline{2^1}$	$\overline{2^2}$	2^3

regulation. Measure the voltage across the transistor—there should be a large ac component. Pick a value for the resistor so that when the voltage is at an instantaneous minimum, the current through the resistor is about 100 mA. If all this is too much trouble, don't bother. Almost any transistor you choose can handle the power even without the resistor.

Final Thoughts

I have described a fairly simple and relatively cheap way of obtaining accurate frequency calibration. There are some inherent limitations. The main limitation is that the counter doesn't actually measure the transmitter output frequency. Since the assorted conversion oscillators are inevitably crystal controlled, there is little cause to worry about significant errors. When you install the unit, it might be a good idea to set the VFO at 000.00 and to zero beat the transmitter output with your calibrator by adjusting the trimmer capacitors on the conversion crystals.

If you can't build the counter described because your rig doesn't fit the criteria, there is still hope. For example, if it is additive **and** subtractive, you might consider using the bidirectional counters that are now available. (These units are presently more expensive than the 7490s.) Direction of count can be controlled by the bandswitch.

If your rig uses some odd VFO frequency, perhaps with a 455 kHz added or subtracted, you can try several things. One is to mix the VFO with a signal that will subtract the odd number. Another is to preset the counter with the reciprocal of the odd number and to set to 9 all digits more significant than the odd number. I don't know how well these ideas will work, since it was not necessary to try them, but they seem to be theoretically sound.

If you feel as I do, that electronic equipment should have uses comprehensible to the layman, witness an additional characteristic of the digital dial. Take VFO knob firmly in hand and give it a spin. The last two digits of the display will be random numbers! Not wishing to become an "accessory before the fact," I won't suggest what you might use those random numbers for, except perhaps compiling a random number table.

Tips On Using Counters

William L Smith

Regardless of whether you found it in the junk yard, built it from scratch, or paid money for it—if you're fooling with a frequency counter, the first thing you'd better do is find out what the counter will do. The most important things you need to know are the frequency range, the input sensitivity, and the accuracy.

If you have an ex-commercial or military unit, the chances are that all this info is set out in the specs or instructions. Most of these types—even the oldest—will count up to 100 kHz. If the front-end tubes are in good shape and everything is adjusted properly, don't be surprised to find that it works okay up to 20 percent or so above the rated top frequency.

If you don't have the specs, or if you have a homebrew counter, you will have to measure the performance yourself. To start out, a hookup something like Fig. 1 is called for. Starting at a low frequency, say, about 1000 Hz, vary the output up and down until you know what signal level is required to make the counter indicate properly. Then, maintaining the output at about twice this level, slowly increase the frequency by tuning the oscillator upward while observing the counter indications. If it is necessary to change bands on the oscillator or signal generator, repeat the output level adjustment to make sure you are above the minimum voltage required for counting. You may find that the signal level needed increases gradually as you tune upwards.

Sooner or later you will reach a frequency at which the counter either fails to indicate at all, or indicates lower as you tune higher. Increase the oscillator output in an attempt to make the unit keep counting. Lower the frequency, if necessary, to make the counter indicate correctly. By now you know with reasonable accuracy the limits of voltage and frequency for normal counting. You should keep these

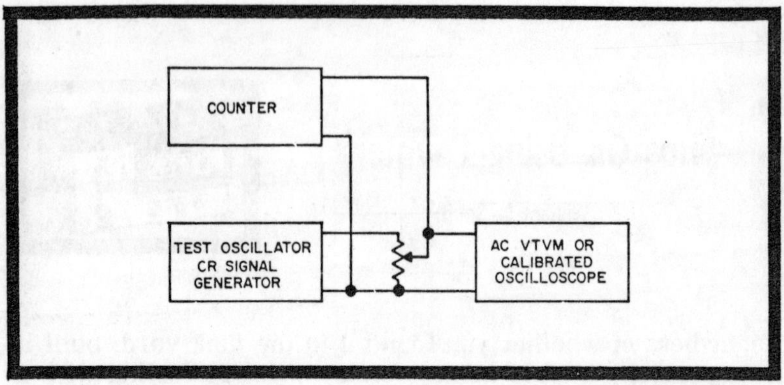

Fig. 1. Measuring the sensitivity and frequency range of a counter.

limitations in mind when designing accessory circuits for your counter.

This is not the place for a long dissertation on oscillator stability, but I think it should be obvious that every counter contains or uses some sort of a frequency standard, and the counter indications are no more accurate than the standard. For example, many older counters use a 100 kHz crystal oscillator operated without an oven and trimmed to "exact" frequency by a big mica compression trimmer capacitor. Such a unit can be no more accurate than the oscillator (which will never put WWV out of business), and its indications will probably be affected by temperature and humidity. Simpler units may use the power line as a frequency standard, and these will be even worse. It has been a few years since I measured the local line frequency on good apparatus. At that time, it was fairly common to find it off a few hundredths of a hertz, but I never found it off as much as one-tenth of a hertz. In other words, the instability was a little under one part in 1000.

Another factor that affects accuracy is the fact that the counter can't split a count. This applies chiefly at low frequencies at which the number of digits indicated is low. For example, if the counter reads **50** (fifty hertz, or fifty millihertz, or fifty **anything**) the true value might be off by as much as one digit; that is, it might be 49 or 51, so the error might be as much as 2 percent. The solution in a case like this is to make the measurement repeatedly and to average the results.

For many jobs, you will be using the counter as an accessory for some other piece of test equipment, to enhance its utility or improve its accuracy. An example is shown in Fig. 2,

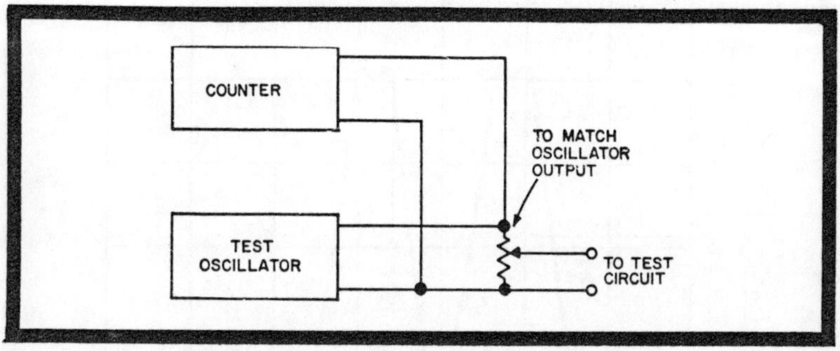

Fig. 2. The counter as a precision dial.

where the counter serves as a precision dial for the test oscillator. The use of an external attenuator or "volume control" allows the oscillator to be run at a high output level (to provide a big signal for the counter) while the level to the circuit under test is adjusted to suit whatever conditions prevail.

MEASURING FILTER RESPONSE

Figure 3 shows the use of this circuit in testing a filter for frequency response. In doing a job like this without a counter, it is usual for the operator to set the oscillator to some even number of hertz or kilohertz, etc., and read the output level. When a counter is used, it is much more convenient to do it the other way. Simply tune the oscillator to a peak, or a null, or to any point of interest on the curve, and read the frequency from the counter. This procedure gives improved accuracy and avoids the possibility of missing any little bumps or valleys in

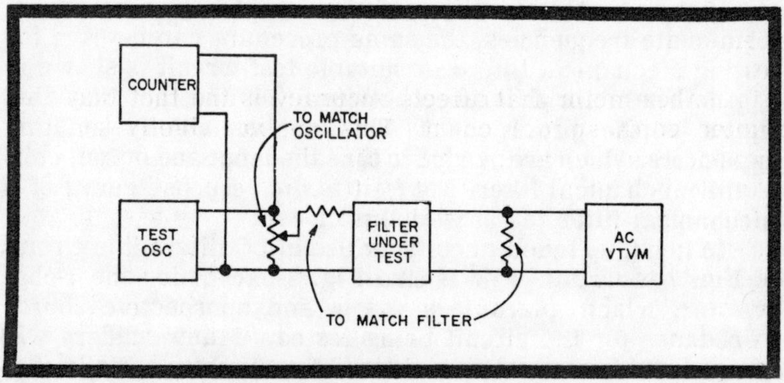

Fig. 3. Measuring filter response.

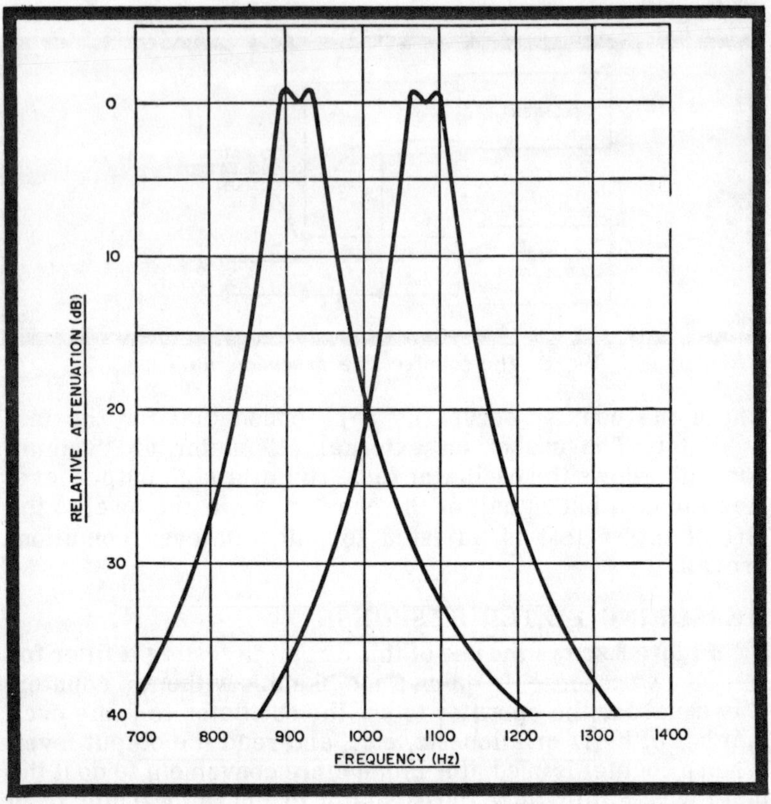

Fig. 4. The response of a pair of audio filters as measured with the circuit of Fig. 3.

the response curve. Figure 4 shows the response of a pair of audio filters built for low-shift RTTY.

If the counter and the oscillator operate at receiver intermediate frequencies, the same procedure can be used for testing mechanical filters. A suitable test circuit is shown in Fig. 5. The main difference between this and the preceding figure is the provision of the two variable resonating capacitors, which are needed to tune the input and output coils of the mechanical filter. Figure 6 is the response curve of a mechanical filter taken with this circuit.

In lining up tuned circuits for use in LC filters, the circuit of Fig. 7 is useful. This is like Fig. 2 except for the 1-ohm resistor, which guarantees a low and nonreactive source impedance for the circuit being tested. (Many readers will recognize this as the basic circuit of a Q-meter.) While it is

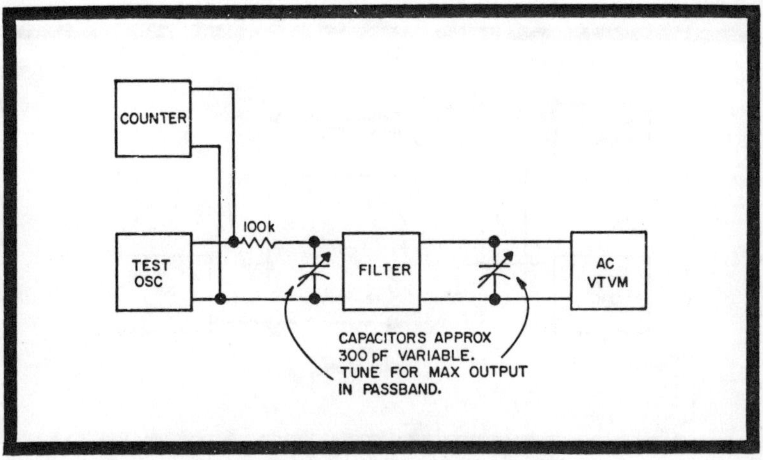

Fig. 5. Testing a mechanical filter.

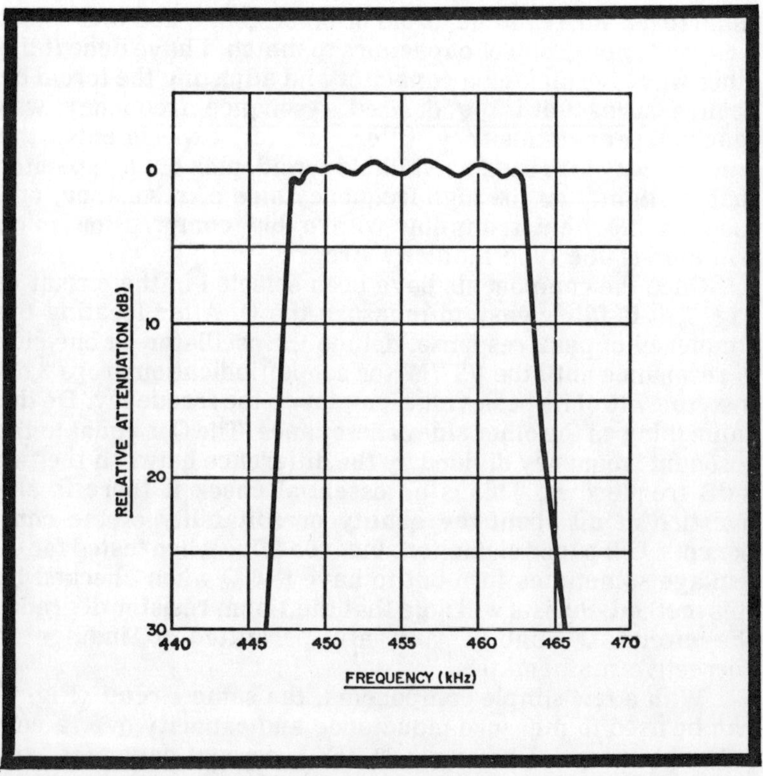

Fig. 6. Response of a wideband mechanical filter, measured as in Fig. 5.

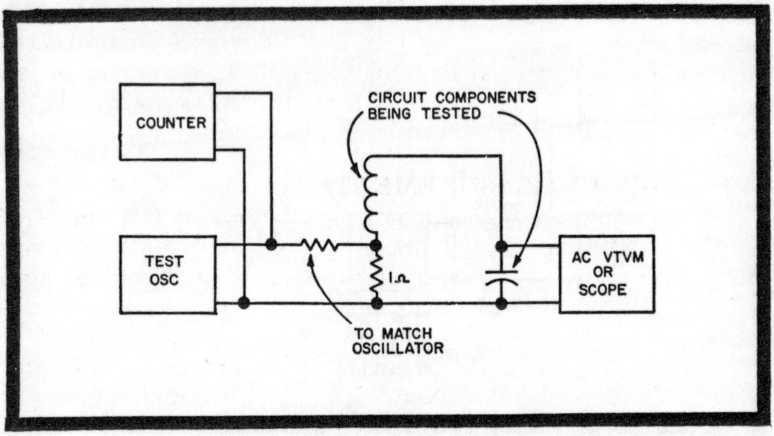

Fig. 7. Checking filter components.

usual to use an available toroid or other high-Q coil and select a capacitor or group of capacitors to match, I have done it the other way; i.e. picking a capacitor and adjusting the toroid by peeling turns until the desired resonance frequency was reached. For temporary filters or for experiments, the quickest way is to use an available toroid, pick out a capacitor that is slightly on the high-frequency side of resonance, and then do the final trimming with a big compression mica trimmer of the type made by Arco.

Once the components have been selected in the circuit of Fig. 7, it is fairly easy to measure the Q. After locating the frequency of peak response, detune the oscillator on one side of resonance until the VTVM (or scope) indication drops 3 dB (i.e., to 7 / 10 of its peak value), and note the frequency. Do the same thing on the other side of resonance. The Q is equal to the resonant frequency divided by the difference between the two 3 dB frequencies. This is an essential check if there is any question at all about the quality or suitability of the components. Old paper capacitors that read fine when tested for dc leakage sometimes turn out to have low Q when checked by this method. Purists will note that the 1-ohm resistor degrades the circuit Q slightly, and are permitted to indulge in corrective mathematics.

With a few simple components, the same circuit (Fig. 7) can be used to measure inductance and capacity over a considerable range. I have a 0.02 uF, 1 percent capacitor, and what I think is a good 100 mH toroid. After I peak the circuit with them, I am ready to substitute or to add other capacitors in parallel, note the frequency change and compute the value of the added component by the usual handbook formulas. Of course, other coils can be substituted, and unknowns of low impedance (low inductance or high capacitance) can be inserted in series rather than in parallel.

HF AND VHF MEASUREMENTS

It is time to consider how to make precise HF and VHF measurements with a low-frequency counter. It takes some more gear, and the first thing you need is a **frequency standard** (and possibly a divider chain). The one I use is shown in block form in Fig. 8. Details of the circuit seem out of place here, as something suitable can be found in almost any handbook. In may case, the standard frequency oscillator runs at 1 MHz, and a divider chain is required to get the standard markers close enough to permit interpolation with a low-frequency counter. If your oscillator operates at 100 kHz or lower, you may not need the dividers. If you have to build this from scratch, it pays to consider transistors, as they eliminate most aging problems and make the power consumption so low you can let the whole works run continuously. My **dividers** are simple, locked multivibrators; they work satisfactorily, and have stayed locked for years without adjustment. My divider chain goes all the way down to 500 Hz, but the most used output is at 50 kHz.

The output of the divider chain at 50 kHz drives a harmonic generator and modulated amplifier, as shown in Fig. 9. This circuit produces markers at 50 kHz intervals up to several hundred megahertz.

The output of the circuit discussed above may be visualized as shown in Fig. 10. Now, if the modulated amplifier of Fig. 9 is driven with the low-frequency oscillator of Fig. 2, each of the 50 kHz markers will grow a pair of sidebands, as shown in Fig. 11. The separation of each sideband from its

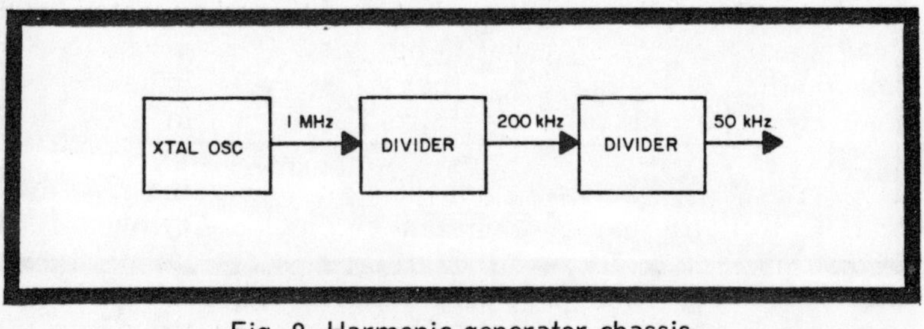

Fig. 9. Harmonic generator chassis.

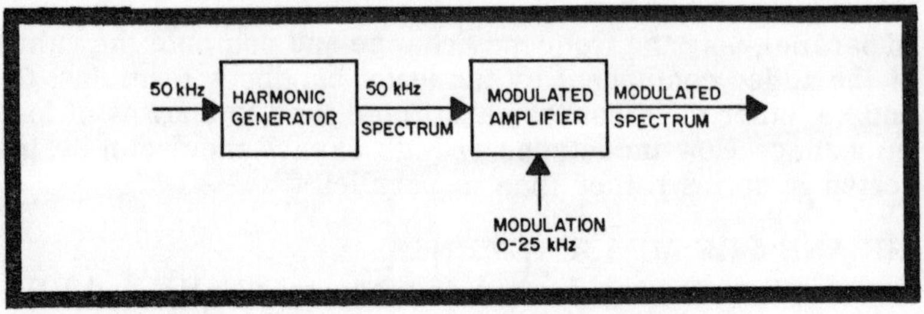

Fig. 10. Output of Fig. 9 without modulation.

parent carrier is equal to the setting of the low-frequency oscillator.

Now we have a scheme for zero beating any signal that turns up in the band.

For example, assume that a signal appears at 144,161 kHz in the 2M band. The nearest marker is at 144,150. From the receiver dial reading, we can tell that the signal to be measured is about 10 kHz above this, or near 144,160. So we set the low-frequency oscillator near 10 kHz, turn up the output, and rock the oscillator tuning until we hear the sideband come close to the signal being measured. Now we can adjust the oscillator frequency very slowly and carefully until zero beat is achieved. (In doing this, it is helpful to adjust the oscillator level—and thereby the sideband level—to obtain the clearest beat.) When zero beat is obtained, we read the frequency of the LF oscillator from the counter. The frequency in this case would be 11 kHz. The frequency of the signal being measured is the sum of the marker, 144, 150, and the modulator, 11 kHz, or 144, 161 kHz.

In using this method you must take care not to get confused. You must be absolutely certain which marker is the nearest, you must know whether the unknown signal is above or below the marker, you must make sure that you use the

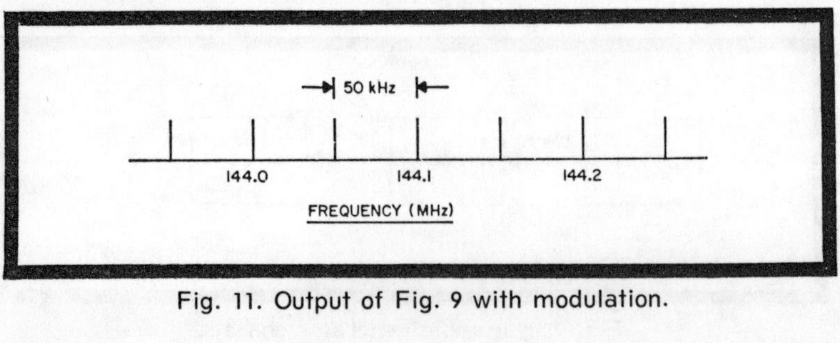

Fig. 11. Output of Fig. 9 with modulation.

lowest possible frequency from the low-frequency modulating oscillator, and you must not overdrive the modulated stage. Obviously the better the receiver dial calibration is to start with, the less the chance of getting mixed up. I have found it a good practice to write down the approximate frequency read from the dial right at the beginning, to make sure I didn't add when I should have subtracted, etc.

MEASUREMENT ERRORS

In making precise measurements, it is necessary to know what errors are in the measuring equipment. The error, if any, in the standard frequency oscillator can be determined by comparing one of its harmonics with WWV. If there is any question about whether the standard oscillator is high or low, it may be resolved by twisting the frequency-adjusting trimmer. If frequent adjustments are considered undesirable, it may be simpler to adopt a policy of always running the oscillator off by a hertz or two at 15 MHz. This is not enough to seriously affect its use as a calibrator, but if it is always run off the same side, the corrections will always be in the same direction. A second but less serious source of error is the error in the counter, or in its internal frequency standard. It is less serious because the portion of the total frequency synthesized in the zero-beating process that is contributed by the LF oscillator is very small. The simplest way to measure the counter error is to measure the marker frequency (50 kHz in my case) on the counter. If the marker reads high, it is a sign the counter is reading high (its internal oscillator is low), and all counter readings must be reduced accordingly.

A good overall procedure to reduce error is the following:

1. Check the standard oscillator against WWV and determine the error.
2. Check the counter against the standard oscillator.
3. Measure the unknown frequency.

a. Note the approximate frequency from the receiver dial.

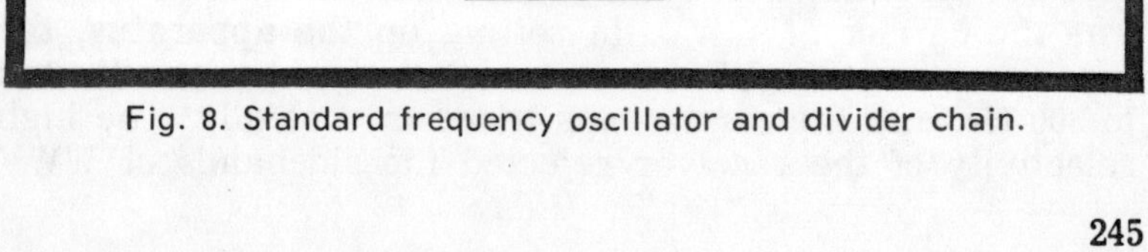

Fig. 8. Standard frequency oscillator and divider chain.

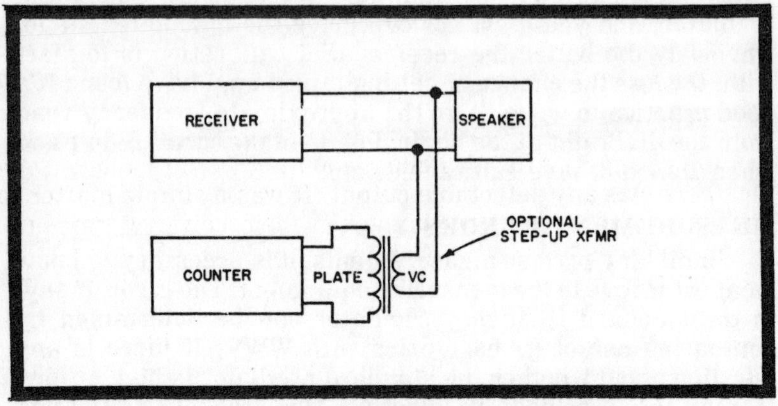

Fig. 12. Counter connected to receiver output.

b. Note the nearest marker, and whether the unknown is above or below the marker.
c. Estimate the frequency for the LF oscillator.
d. Turn up the LF oscillator level and adjust for zero beat.
e. Read the LF oscillator frequency from the counter.
f. Repeat a, b, c, and d several times, if possible.

4. Repeat 1.
5. Repeat 2.
6. Correct the indicated frequencies for the apparatus errors.

There are a number of tests that can be made with the counter connected to the output of a receiver. In making this connection it is a good idea to use the high-impedance or 6-ohm output of the receiver, if it has one, so that the counter will get a signal of adequate level without overdriving the speaker or headset. If there is no high-impedance output circuit and the counter is insensitive, it may be necessary to use a stepup transformer between the speaker circuit and the counter input. An ordinary plate-to-voice coil transformer connected backwards works fine (see Fig. 12).

If a strong signal can be obtained from a standard frequency station or any other source known to be stable, it can be used to measure the warmup drift of the receiver. Figure 13 shows a curve taken in this manner on a commercial receiver operating at 15 MHz. In this case, the standard signal was the carrier of WWV. In setting up the apparatus, the receiver was set for CW reception with the bandwidth adjusted to 500 Hz, and the carrier was tuned in carefully. The high selectivity of the receiver rejected the sidebands of WWV (ticks and tones), and the receiver output was a single clear beat note of about 900 Hz. After the initial adjustment, which required about 5 min, the set was turned off for 4 hr so as to stabilize in the cold condition.

The counter was warmed up for a few hours before turning on the receiver. After switchon, counting was started as soon as there was any detectable output. It was a simple matter to write down the frequency changes as the receiver warmed up. Figure 14 shows that the "peak-to-peak" drift was only 110 Hz. After warmup, the residual drift was less than 10 Hz per hour.

With a receiver of known stability, it becomes practical to use it and the counter to measure the drift of other signal sources. With the bfo turned on and the receiver selectivity slacked off a bit, the combination may be used to measure the shift of a RTTY transmitter. To do this, it is necessary to put the transmitter alternately on steady mark and steady space, for several seconds each. In most RTTY setups, this can be done by opening the break switch on the keyboard. Several readings should be taken on both mark and space, and each group averaged so as to eliminate any drift in either transmitter or receiver.

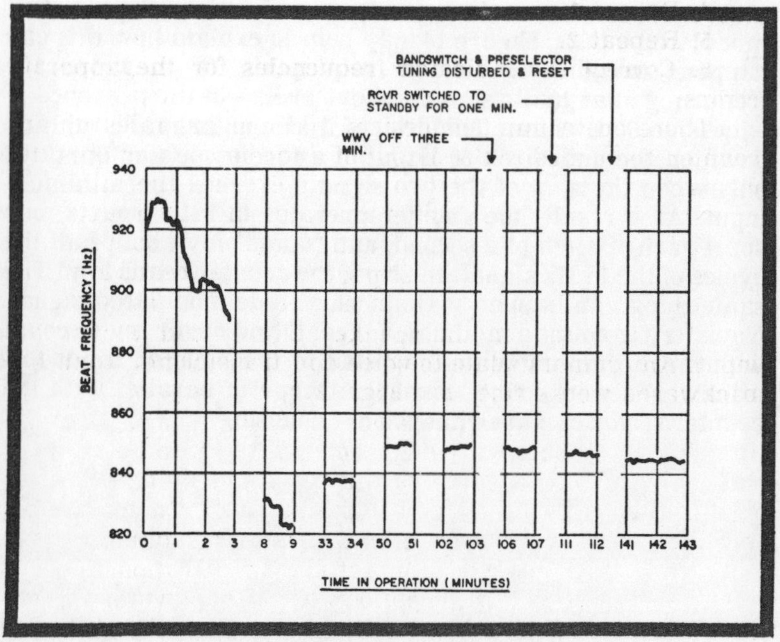

Fig. 13. Receiver warmup drift.

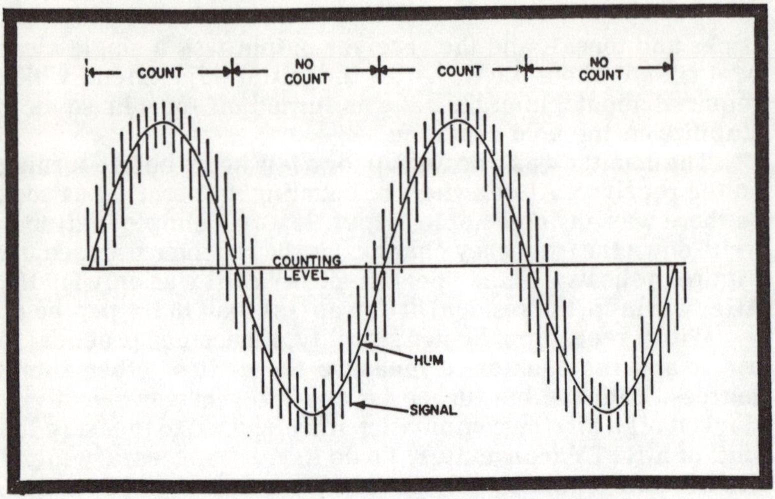

Fig. 14. Incorrect counting with multiple inputs.

Counters don't always tell the truth. Some can be made to lie by changing or subnormal line voltage, high temperature, out-of-tolerance tubes, etc. But there are times when even the best units lie. In experimental hookups the most frequent causes of erroneous indications are multiply inputs and modulated inputs. Figure 14 may help to explain how this can happen. Imagine a counter attempting to measure the frequency of an audio signal—about 1 kHz—in the presence of excessive 60 Hz hum. The desired 1 kHz signal rides up and down on the undesired 60 Hz hum, and the counter operates only when the sum of the two signals exceeds the minimum input. As a result, the counter operates in little spurts, one spurt each sixtieth of a second, and counts only about half the cycles of the 1 kHz signal. In short, the counter reads low! The same thing can happen with an amplitude-modulated signal. Never try to count a modulated signal, and never try to count with more than one signal applied to the counter input terminals. When in doubt, connect a scope in parallel with the counter so you can see what you are counting.